Academic Press Rapid Manuscript Reproduction

FISHERY DEVELOPMENT

FISHERY DEVELOPMENT

WILLIAM F. ROYCE
College of Ocean and Fishery Sciences
University of Washington
Seattle, Washington

1987

ACADEMIC PRESS, INC.
Harcourt Brace Jovanovich, Publishers
Orlando San Diego New York Austin
Boston London Sydney Tokyo Toronto

ACADEMIC PRESS, INC.
Orlando, Florida 32887

United Kingdom Edition published by
ACADEMIC PRESS INC. (LONDON) LTD.
24–28 Oval Road, London NW1 7DX

Library of Congress Cataloging in Publication Data

Royce, William F.
Fishery development.

Includes index.
1. Fishery management. 2. Fishery conservation.
I. Title.
SH328.R74 1987 333.95'615 86-47988
ISBN 0–12–600955–4 (alk. paper)

PRINTED IN THE UNITED STATES OF AMERICA

87 88 89 90 9 8 7 6 5 4 3 2 1

CONTENTS

III FUTURE STRATEGY

PREFACE

Fishery development is widely practiced yet ill defined. It is frequently considered as a part of, or an addition to, fishery management, which is usually defined as the practice of governmental control of fishing with the primary goal of conserving the resources.

But fishery development is clearly a part of economic development that is fostered by all governments. As such an activity, its goal is to increase the social benefits from the fisheries which, in the long term, must include their conservation. Therefore, fishery development is a more inclusive concept and a much more difficult goal than conservation.

Regardless of its definition, fishery development is practiced by most governments, the poorer ones with assistance from international development agencies. It is becoming much more widespread despite the peaking of production from the world's major fish stocks. And it is commonly undertaken by governments unfamiliar with the problems, unprepared for the complexities, and expecting major benefits from short-term projects.

The troubles are manifold. The fisheries are usually regarded as a unit, yet they are an extraordinarily diverse assortment of activities. Most of them depend on resources of the commons, which must have skillful government control if social values are to be maximized. Such control depends upon acceptance by the fishery communities, and requires a viable blend of politicians, administrators, and scientific experts for its execution. The difficulties are common to both developed and less developed countries because the traditional coastal fishing communities in both are subject to the unrelenting pressure to modernize.

This book provides an overview of the whole process of fishery development and an appraisal of how it can be done more efficiently. My objectives are to identify the technical, social, political, organizational, and time requirements of long-term development programs. I have emphasized methods for increasing the long-term social benefits from the fisheries, including more income for poor fishers or fish farmers, more employment, more highly desirable protein food for local markets, more recreation where fish are important for that purpose, and more export earnings. The audience will encompass a broad spectrum of people who are concerned about the fishery development decisions being made by both industry and governments. Planners and administrators will find helpful information for blending their fishery development projects into long-term programs. Fishery students also should find it useful as a text for a general course.

Such a mixture of investment, technical, and social problems in the fisheries has been little studied, except perhaps by a few large fish businesses for their internal use, so my approach will be more pragmatic than analytic. I shall summarize the structure and trends in the fisheries, review numerous case histories of fishery development, and analyze the factors responsible for successful approaches. Special emphasis will be given to the formation of government policy with respect to the market fishing because, as with use of all resources of the commons, government must be the controlling partner.

This broad perspective of fishery development has been much easier to prepare because of the excellent publications on experience with economic development, and technical information about the world's fisheries, which have been collected and published by United Nations Organizations. The World Bank, with more than three decades of experience in assisting economic development, has discovered the difficulties. The Fisheries Department of the Food and Agriculture Organization of the United Nations, with its technical experts from many nations, has become the outstanding fishery consultancy in the world.

I am indebted to Mary C. Beaussart of the International Development Research Centre in Ottawa, and to Milan J. Kravanja of the Office of International Fisheries, National Marine Fisheries Service in Washington, who provided or located many references. Thanks are due also to David L. Fluharty, Orra B. Kerns, William J. McNeil, Lynwood S. Smith, Robert R. Stickney, and Richard B. Thompson, who critically reviewed all or parts of the manuscript. I am especially indebted to Roy I. Jackson, who critiqued the entire manuscript, and with whom I have had many hours of discussion about fishery development. None of them, however, bears any responsibility for errors, and they do not necessarily share my opinions.

William F. Royce

ACRONYMS AND INITIALS

ADCP	Aquacultural Development and Coordination Programme (FAO)
AFPA	American Fisheries Promotion Act
AID	Agency for International Development (U.S.)
BOBP	Bay of Bengal Program (FAO)
CAG	Compañía Administradora de Guano (Peru)
CECAF	Committee for the Eastern Central Atlantic Fisheries (FAO)
CGIAR	Consultative Group on International Agricultural Research
CIDA	Canadian International Development Agency
CIPS	Confédération Internationale de la Pêche Sportive
DANIDA	Danish International Development Agency
EEZ	Exclusive Economic Zone
EIA	Environmental Impact Assessment
EIFAC	European Inland Fisheries Advisory Commission (FAO)
EIS	Environmental Impact Statement
EPCHAP	Empresa Pública de Commercialización de Harina y Aceite de Pescado (Peru)
EPSEP	Empresa Pública de Servicios Pesqueros del Perú
FAO	Food and Agriculture Organization of the United Nations
FCMA	Fisheries Conservation and Management Act (U.S.)
GDP	Gross Domestic Product
ICES	International Council for the Exploration of the Sea
ICLARM	International Center for Living Aquatic Resources Management (Manila)
ICNAF	International Commission for the Northwest Atlantic Fisheries
IDRC	International Development Research Centre (Canada)

IMARPE	Instituto del Mar de Perú (Peru)
ITQ	Individual Transferable Quota
LOS	Law of the Sea
MSY	Maximum Sustainable Yield
NEPA	National Environmental Policy Act of 1969 (U.S.)
NMFS	National Marine Fisheries Service (U.S.)
NOAA	National Oceanic and Atmospheric Administration (U.S.)
NORAD	Norwegian Agency for Development
OPEC	Oil Producing and Exporting Countries
OSY	Optimum Sustainable Yield
PESCA-PERU	Empresa Pública para la Elaboración de Harina y Aceite de Pescado (Peru)
SCSP	South China Sea Program (FAO)
SIDA	Swedish International Development Authority
S-K	Saltonstall-Kennedy Act of July 1, 1954 (U.S.)
TAC	Total Allowable Catch
TCDC	Technical Cooperation among Developing Countries (FAO)
UNCLOS	International Conference on the Law of the Sea
UNDP	United Nations Development Programme

PART I

LESSONS FROM THE PAST

The best way to plan future strategy is not merely to project the trends that have shaped the present, but to understand them. The world's fisheries have developed more in the past three decades than in all previous history, and changes are happening ever more rapidly.

This section is a brief description of the fisheries and their role in development. It starts with an overview of economic development and continues with a summary of the fisheries, including their practices, production, and management. It is preparation for the subsequent description of fishery development experience and an evaluation of fishery development problems.

Chapter 1

Issues and Challenges

Fishing[1], that ancient human activity, has developed throughout the centuries until now, almost surely, all of the wild living resources of the waters have been discovered, and the total production of most conventional kinds has peaked. Yet we want to develop them more!

Almost every country and international assistance agency has fishery development programs. Interest in them erupted during the long discussions of a new Law of the Sea (LOS) and, after a consensus was reached in 1974, culminated in acceptance of the concept of a 200-mile Exclusive Economic Zone (EEZ). This provided new authority over the marine fisheries and occurred after decades of extraordinary growth in world fishery production. These two occurrences led most coastal countries to expect major development opportunities in their fisheries.

The expectations were politically reinforced in 1981 when the Food and Agriculture Organization of the United Nations (FAO) asserted that "A revolution has occurred in the potential of fisheries to contribute to a new international order," and stated its "...intention to take a lead...by helping the developing countries to secure their rightful place in world fisheries."[2] The FAO followed up with extensive preparation for a World Conference on Fisheries Management and Development that was held in Rome, June 27 - July 6, 1984.[3]

If the total production of most resources has peaked, why is there so much interest in further development? The reasons vary from place to place, but some are that the fisheries provide employment, a desirable food, an important recreation, a contribution to gross national product, and/or a source of foreign exchange earnings. These reasons may be strengthened because of potential social benefits for poor people. The subsistence fisheries provide essential food for rural poor in remote communities. Many small-scale commercial fisheries are only marginally profitable and need assistance from governments. Other reasons include the widespread interest in recreational fishing, especially in developed countries, and the political pressures to increase it.

These perceptions have led to the fundamental dichotomy in fishery management. Use of the fish of the commons has been regarded for millenia as a fundamental right of all people. Exercise of this right has led traditional fisheries, through local preferences or government action, to forego excess ingenuity in their competition for the resource. But the modern market fisheries, which must compete for a share of the resource as well as a share of the market, depend upon constantly increasing ingenuity in fishing, processing, and market systems.

1.1 A NEW ERA IN THE FISHERIES

The old era is ending after two unique and profound changes in the fisheries. First was the great expansion of the world's ocean fisheries in only two decades, the 1950s and 1960s, when production tripled. This expansion was a feat of the industrial sectors of the larger nations with virtually no restraint by governments. It involved innovations in powered vessels, electronic fish-finding equipment, synthetic netting materials, sophisticated preservation and processing equipment, distant-water fishing, and worldwide marketing; all by organizations able to marshall financing, information, and skills on a scale unprecedented in the fisheries.

The hopes for greater benefits from the fisheries that followed this expansion generated the second event. The developing countries, which line the shores of most of the oceans, hoped that they could have a larger share in the use of the fishery resources. They vigorously supported new international laws that permit jurisdiction by each coastal country over the fishery resources up to 200 miles from shore, a zone that contains more than 95% of the world's conventional marine fishery resources.

These laws have caused and will continue to cause profound changes in the ocean fisheries of the world. The distant-water fisheries of the world's leading fishing nations were mostly off the coasts of other countries. They are no longer legally possible without agreement with coastal countries which in many cases, will bring such activities under government control for the first time. The control will be exercised by the coastal country, which will be in a position to negotiate an agreement with benefits to itself.

The coastal country will have an opportunity to divide the catches between domestic and foreign fishing, to protect its fish resources for its own fishers, to phase in its own large-scale fishing, and to increase its production of fish. If the coastal country is a developing country, as most of them are, it will have an opportunity to help its small-scale fishers and obtain additional food for poor people. If it is a developed country, it will have the opportunity to protect its large-scale fisheries, reduce its imports or increase its exports of fish, and perhaps collaborate in joint fishing ventures with other countries.

Thus the new era in fisheries finds countries grappling with the radical redistribution of authority over the limited commercial fisheries. But in addition, the expansion of two other kinds of fisheries demands attention.

First are the recreational fisheries. These provide a popular outdoor recreation in the developed countries and a few of the developing countries. In ancient times fishing was a sport of kings and the wealthy, but now it attracts widespread participation. In many countries the recreational fishers, who are far more numerous than the commercial fishers, can obtain exclusive allocation of some freshwater and coastal fisheries to themselves.

Second are the culture fisheries, the fish farming. This ancient activity is becoming the principal source of increased fish production. A small fraction is public fish farming, which enhances the wild stocks of public waters, but the majority is private fish farming to provide food for markets, fun for anglers, and delights for aquarists.

1.2 THE DIVERSITY OF THE FISHERIES

The fisheries are usually only a small part of national activity in comparison to agriculture and industry, and a minor concern of the institutions of government, but the fisheries have a unique relationship to each of these. Rural fisheries may be closely related to agriculture socially, because many poorer fishers live in villages, where some are part-time farmers, and fish are commonly distributed along with other foods in the markets. In addition, some farmers may raise crops of fish. Other commercial fisheries may be related to industry through a sophisticated use of capital for large-scale fishing in distant oceans, for complex processing, and for international marketing of the processed food.

Fisheries are based on complex resources which include hundreds of kinds of fish Each species has its own habits, living in different kinds of water, and has different market qualities, but together they provide excellent protein-rich food in far greater variety than animal agriculture. The source of about 85% of the supply is the wild stocks of the oceans and fresh waters, where the fish are public resources of the commons. The other portion, which is rapidly growing, is fish farming, a relatively sophisticated kind of agriculture.

Even greater diversity occurs in the business of the fisheries. Fishers who seek food have vastly different kinds of equipment ranging from a simple line or trap for the subsistence fisher to a fleet capable of fishing any ocean in the world. Their production per person varies at least a thousandfold, as does their individual income. Their products, which can be consumed locally within a day or two without preservation, require skilled processing and handling before reaching distant consumers. The merchants buying and selling fishery products range from remote village people remaining in their own communities to transnational corporations.

Some products feed the poorest people, others entice the richest, and about one-third of current world production now enters international trade. Fish consumption also has some special social aspects. Some groups of people refuse to eat fish of any kind; others have objections to certain species such as those without scales. Still other people may consume fish at certain times for religious reasons.

Another social dimension is growing rapidly, especially in developed countries, where large numbers of fishers seek outdoor recreation in fishing. When these people can have a pleasant experience in a clean environment, the fun is more valuable than the food.

1.3 THE GOVERNMENT BURDEN

The desire for economic development has put governments of developing countries in a new and challenging role. The economies of such countries have typically been largely rural and agricultural. Their governments are attempting to do much more than it is customary for governments of the advanced countries to do, which is even much more than the governments of those advanced countries did at early stages of their development.(4)

Unfortunately, most of the fisheries over which the coastal countries have gained authority are examples of the inevitable result of free commercial fishing,(5) "Fisheries that are unlimited become unprofitable". Experience worldwide has shown that commercial fisheries eventually become unprofitable when the fishing effort is unlimited and remain unprofitable as long as the fishing effort is freely competitive. The developed countries, that discovered and applied the scientific principles of fishery management, have failed to keep profitable most of their commercial fisheries. New policies must be found and applied if the commercial fisheries are to contribute to economic development. It will not be easy.

But the failures of commercial fishery management must not be blamed on the science which is essential for either fishery management or development. The science is required for determination of the nature and sustainable yield of each stock. A stock is a unit of the resource, preferably a single race of a species, that can be fished separately from other units. The nature of the stock is its location, migratory habits, and environmental requirements. The sustainable yield is the result of the balance of birth, growth, and death rates at different population sizes.

In addition to such biological information on the resources, a fishery manager needs reliable and continuing information on the amount of fishing effort by different kinds of gear and on the quantities caught categorized by species and size in order to estimate the impact of fishing on the stock.

Such information is the basis for fishery management that has the principal objectives of conservation of the resources and of obtaining a desired total physical yield from each stock. The management typically specifies measures to prevent waste of fish and a quota for the amount that can be caught without regard to the economic state of the fishery.

Fishery management of this kind requires policy formulation by government that establishes the necessary laws and government organizations able to obtain the scientific information, make regulatory decisions, and enforce the decisions.

Fishery development is a much broader function, requiring government support of the fishery management functions plus intervention in the fisheries in order to maximize the socioeconomic benefits. Whereas fishery management is a reactive function based on the initiatives of the fishery sector, fishery development is a proactive function intended to stimulate change in the fisheries. It is a combination of the scientific task of providing an information base, the political task of formulating new policies, and the long-term administrative task of maintaining a coherent program. It is a complex challenge because of the diversity of the people engaged in the various production systems, the limitations of fish stocks, and the competition for the use of the waters.

Few other human activities span such diverse operations. Subsistence and small-scale fishers pursue an ancient practice still akin to hunting and gathering. Their simple ways persist in many countries among millions of poor people who have few other opportunities. Yet they may compete with industrialized fisheries that use sophisticated equipment to provide the great majority of total production.

Still another dimension of fishery development is its dependence on suitable water. Water, like most of the fish stocks, is also a public resource of the commons. Fresh water is limited in quantity, scarce in most localities, and commonly wasted or abused. Salt water is abundant in the oceans, but in the coastal zones where most of the world's fish stocks occur, it, too, is commonly abused. Management of the water quality is also a combination of scientific, political, and administrative tasks, which are usually even more diverse than fishery development because of different interests in and the great importance of the water.

Fishery development, like any economic development, requires effective and efficient use of the available skills, capital, and resources. The skills and the capital must be balanced between private and public sectors, whereas management of the resources is traditionally a public responsibility. Thus the private and public roles are interdependent to a greater extent than in the agriculture or industry of almost all countries except those with centrally planned economies. The predominantly private roles of catching, processing, and marketing fish must be kept as efficient and equitable as possible, through public and private teamwork, to maintain their unique capacity for flexibility and entrepreneurship not typical of the public sector. The predominantly public roles of preventing overfishing and allocating the resources of both water and fish among competing users must be performed with expert knowledge and skill to balance the many interests and promote growth with both conservation and equity (see Section 10.3).

A common goal of fishery development is improvement of the small-scale fisheries through government intervention, frequently with international assistance to less developed countries. The programs are chosen because the fishers are poorer than average, their catches are important food in the localities near where they live, and the fish resources seem to be capable of producing more. Such a program in lesser developed countries is often part of a comprehensive rural development program. But even in developed countries, many traditional fishing communities have failed to develop as rapidly as average communities, so subsidized assistance to them is arranged.

The extended authority of coastal countries under the LOS adds further diversity to the government problems of fishery development because every country must agree with neighboring countries on boundary lines in the sea if the area of coastal jurisdiction is to be recognized. Further, the management of coastal resources, like the management of riverine resources which move across political boundaries, must be agreed upon by neighbors if it is to be effective. Thus, no nation alone can act effectively on its fishery problems within its boundary waters, but rather it must act in concert with its neighbors in order to balance each other's interests in both water and fisheries.

These problems of fishery development typically have been approached from the specialized viewpoints of fishery scientists, economists, cultural anthropologists, political scientists, lawyers, fishers, or fishery businessmen. The scientific and technical inputs have been essential but not adequate for the fundamental goal of effecting social change. Achieving this goal depends on an evolution of human values among the people affected; a process in which those people must make decisions.(6) Few fishery development programs have accomplished their long-term goals, which usually require many years to attain.

The goals seem to be so simple and desirable, yet too often setting them neglects a hard examination of the feasibility and implementation of the steps needed to achieve them. A central problem is enhancement of the skill of government people and organizations, especially because of division of responsibility among ministries of the central government and between central and local government. Another is the tendency to consider the fisheries separately from agriculture, industry, and community development, all of which have an impact on the fisheries.

Perhaps the most difficult problem is overcoming the inherent resistance of people to change, and changes are happening ever more rapidly. Fishers are constantly improving their equipment--regardless of the state of the resources. Processing methods and products are improving with more attention to distant markets. A few transnational corporations with major economic power over the fish markets are providing quality fish at low cost in many countries.

The new policies will require extraordinary skill on the part of government because the technical, social, and business problems of the fisheries are complex. Governments will need the scientific ability to assess the sustainable yields of the resources, the legal ability to promulgate effective regulation of the fishing, the diplomatic ability to negotiate with neighboring countries over the use of resources in rivers and oceans that cross international boundaries, the political sensitivity to guide changes in the lives and livelihoods of rural poor in order to restrain the traditional right of piscary, the business sensitivity to encourage private organizations that can perform many functions better than governments, and the organizational ability to manage dynamic changes in the fisheries over a long period of time.

NOTES AND REFERENCES

1. Fish, fishing, and fisheries are used in their collective sense to include all animals, especially finfish, mollusks, and crustaceans, as well as the plants harvested by the fisheries.

2. FAO, "Fisheries Development in the 1980's." FAO, Rome, 3 pp. (brochure), 1984.
3. FAO, "Report of the FAO World Conference on Fisheries Management and Development, Rome, 27 June - 6 July 1984." FAO, Rome, 60 pp. + appendices, 1984.
4. E. S. Mason, The planning of development. *Sci. Am.*, *209*(3):235-244 (1963). This entire issue is devoted to technology and economic development as seen about 1960.

5. M. Graham, "The Fish Gate." New Ed. Faber and Faber, London, 199 pp., 1949.
6. E. S. Dunn, Jr., "Economic and Social Development: A Process of Social Learning." Johns Hopkins Press, Baltimore, Maryland, 327 pp., 1971. See Chapter 5, the Role of Social Goals and Values, pp. 160-185.

Chapter 2

Economic Development

Fishery development has been a part of efforts by governments to enhance overall development for several decades. Most of the problems encountered in fishery development have counterparts in other sectors, where the experience provides useful examples to fishery experts.

Development is an ill-defined process of improving the living conditions of people and correcting the disparities among them. Development continues, of course, through the traditional operations of the marketplace, but this leaves inequities which governments seek to overcome. The governments of all nations strive to speed up their development and to assist their poorer people by borrowing money or by diverting resources from the well-to-do in order to invest in future growth. And among nations, the developed countries, for reasons of their own, want to help the developing or less developed countries. Much of this effort is focused on on promoting trade and alleviating poverty.

2.1 GOALS AND POLICIES

The goal of economic development is of the broadest kind--to improve the well-being of entire populations of people. As such it is a function, indeed an imperative of national governments. Although it is commonly described in terms of economic growth, it must promote social equity, at least in the long term, as a part of the process.

The strategies for development, as should be expected with a generalized and new activity, are diverse and controversial. An overriding goal in every country is to help the poor, especially in the less developed countries where the poor are a large fraction of the population. Every development planner is challenged by the need to find the combination of economic, social, scientific, technical, and organizational changes that will accelerate the rate of development. And every development planner has been dismayed by the difficulty of executing the best laid plans. Consequently, it is possible within the limited scope of this book merely to identify prob- lems that should be expected.

The problems arise throughout the full range of development activities, from the motives of the donor through the intricacies of policies, projects, and programs, to the motivation and means of recipients to change their activities. The strategy of the planner or manager of development needs to be considered in its broadest sense if it is to encompass the problems, which it must do to be successful.

The description of development as economic may, in fact, be a misnomer because all enhanced development is political in the sense that it must satisfy the value judgments of people. But it must also have a sound basis in the facts about each situation. This poses a special challenge for all developers who must balance a mixture of facts and values.

The political dimensions range through all levels of government to the United Nations, where many of the lesser developed countries have sought cohesion in a "Group of 77" (now about 130 countries). These countries adopted a declaration and a program of action about a new economic order at the United Nations General Assembly session in 1974. They would replace an economic order, which they characterize by inequity, domination, dependence, narrow self-interest, and segmentation, with a new political order based on equity, equality of sovereign nations, interdependence, common interest, and cooperation. In effect they held the developed countries responsible for their plight. They have continued in this direction, and a recent spokesman stated, "All concerned must listen to the voices of the poor, who have paid the highest price for the passing order and can no longer be kept in convenient silence." He also advanced a definition of development as "...the continuing process of the liberation of peoples and societies. There is development when they are able to assert their autonomy and, in self-reliance, to carry out the activities of interest to them."[(1)]

The world economy at the end of the 1970s was functioning so badly that it was seen by some as dangerous to the entire world community.[(2)] The problems were recognized as being very complex and the situation so urgent that an emergency program was recommended to include, (1) a large scale transfer of resources to developing countries, (2) an international energy strategy, (3) a global food program, and (4) a start on major reforms in the international economic system.

But many aspects of the economic order are seen differently from a less political viewpoint. Morawetz[(3)] pointed out that the average per-capita income of the developing countries grew more rapidly between 1950 and 1975 than their income or that of the developed countries had ever done in the past. The average, of course, includes a wide range of growth, and some countries, or some peoples, may have become worse off in absolute terms.

Kahn,[(4)] who has analyzed recent development trends, has argued that the widening gap is really a stimulus for transferring the benefits of economic development to the poor. Many countries, once they have risen above the lowest levels, have experienced very rapid growth; for them the relative gap has diminished. Countries with this experience include several in eastern Asia that have fitted the strategies of the developed countries to their own cultures.

A diversity of views frequently exists at the level of the smallest political unit, a rural village. Here the people usually have reached an accommodation with their physical environment and the village has increased in population to a size at which the people live as well as in nearby areas. They probably have also reached a social accommodation among themselves with a central government based on custom, if not law. If the village is a target for economic development, the people will expect a change that they will regard as risky at best. Some will expect benefits; others may expect the benefits to be at their expense and resist the change, perhaps violently.

2.2 A RECENT ACTIVITY

Development, with its present emphasis, started only a few decades ago. The international efforts to intervene in the process began in earnest after World War II, first, with repair of the damage resulting from the war, and second, with assistance to the newly emerging nations, especially the former colonies. The first was largely completed during the 1950s, the second began during the 1950s and continues to this day.

But the first did not supply a model for the second. The repair of the wartime damage in the already developed nations involved the transfer of western practices to countries already accustomed to them. Furthermore, the prevailing economic theories of the time prescribed industrial production as the path to development, which was expected primarily to increase demand for agricultural products and secondarily to expand agricultural production.(5) The assistance to the lesser developed countries involved, in most cases, loan programs with little attention to the difficulties of transferring western practices to the very different cultures in the recipient countries.(6)

Each of the recipient countries is different from the others, but they fall into three cultural groups. Most South American countries have their dominant cultural roots in Spain and Portugal, countries which conquered the native peoples. They were colonies for a time but they declared their independence a century or more ago. Pockets of the native cultures remain, but the dominant culture dates from the large immigration of Spanish and Portuguese settlers in the seventeenth century. This culture, unlike that of the European countries, has had little industrial development and remains dominated by a small number of large landholders.

The central African countries, on the other hand, were colonies, each of which consisted of many tribal groups. The power of the tribal chiefs was broken and colonial identities became established under the European powers. Along with the identity came major cultural changes, such as reduction of tribal migrations, emergence of national languages, land ownership, the concept of a rule of national law, and national administrative systems. Many aspects of European government had been introduced by 1960, the approximate time of independence. But the new governments, with their limited experience and the diverse cultures of their peoples, remain fragile.

A third pattern exists in the Asian countries where the people maintained confidence in their own cultures, some of which have existed for millenia. Here the pattern of development varies greatly among countries, but western ideas and values have been blended with traditional ones in some countries to produce extraordinary development. Other countries, handicapped by instability and regional violence, remain among the poorest.

2.3 THE TIME DIMENSIONS

The urgency of the development needs and the diversity of development tactics commonly lead to unrealistic expectations. Great hopes have been attached to projects that can easily deliver things such as financial, material, research, and advisory assistance; and indeed, loans and advice can be delivered in a month or so. If these first steps toward development lead to new facilities, new infrastructure, or special training for individuals, such things can be accomplished in a year or so. But every such measure, if it is to achieve its ultimate goal of benefiting large numbers of people, leads to changes in people's lives and expectations that must be fostered by governments through changes in policies and organizations. This takes much longer. The experience in rural development (including small-scale fishery development) indicates that the people affected will usually need programs spanning a decade or more to realize significant benefits.

And every sectoral change can be accommodated only by changes in other sectors. For example, the mechanization of fisheries may result in much greater catches and a loss of artisanal employment, while requiring factories for processing and preserving, transportation to new markets, sophisticated technical services to maintain equipment, a shift from a rural to a city seaport base, and skilled government services to maintain the resources. Such social adjustments may involve many sectors of the economy and will require much more time.

The actual pace of development can be realized by comparison of the relative economic development of communities at different times and places. People in many rural communities of the less developed countries live much like the rural people in western Europe did in the early nineteenth century. And the people in some remote rural fishing communities fish and live much like other fishers did two or three millenia ago.

Development is a continuing process to which each bit of advice, money, or equipment can contribute only briefly. Developers must be prepared for programs of indefinite length. The ultimate goals will never be achieved; the intermediate goals will be merely approached. The objectives of projects will need frequent revision according to the changing social circumstances.

2.4 THE CHANGING FOOD AND HUNGER PROBLEMS

The overall adequacy of food supplies has always been a major concern but one that waxes and wanes according to rates of population growth and temporary changes in supply due to weather, wars, or other circumstances. Alarms were sounded in the early 1970s after a series of shortfalls in important crops due to weather, followed by reductions in carryover grains due to storage costs.[7] These concerns were exacerbated by perceptions that populations were increasing faster than food supplies and by rapid increase in food prices.

Since then, overall food production has increased at a rate faster than that of the population, and therefore the supplies in the early 1980s are adequate. An important part of the increase has come from applications of research that have substantially increased production of grains per unit area in many countries (see next section).

The overall statistics on food production[8] have been encouraging. Indices of per-capita production in 1982 were 108 for developing countries and 106 for developed countries (1970-1974 = 100). But disparity among regions remained; the index for Africa had fallen to 92. Nevertheless, an encouraging statistic is the quantity of meat produced that increased from 1975 to 1982 in the devel- oping countries by 32%, and in the developed countries by 13%. The fastest growing part of the meat supply was poultry, which increased in the developing countries by 79%.

But an adequate total supply of food does not eliminate undernutrition, which occurs in all countries and is pervasive among large proportions of the populations in some countries. It is now recognized that overcoming hunger requires employment and income for the hungry, not merely a gift of food.[9] The majority of people threatened by undernutrition are the people of rural areas, the unemployed in the cities, the landless laborers, the marginal farmers, and the poor coastal fishing families. Solution of their problems requires appropriate national policy and effective local institutions.

2.5 SOURCES OF INTERNATIONAL DEVELOPMENT ASSISTANCE

Much of the success of an international economic development program depends on the relationship between the donor and the recipient. The numerous donors have different reasons for supporting development and different practices that profoundly influence the scope, duration, and effectiveness of development programs.

Donors include national governments, charitable foundations or other private organizations, United Nations organizations [especially the United Nations Development Programme (UNDP)], and the International Development Banks. The latter include the World Bank with its three branches and the Asian, African, and Latin American Development Banks. Private banks with international operations extend credit to developing countries or large businesses much as they do in the developed countries.

The motives of donors are fundamentally humanitarian but seldom that alone. If the donors are governments, the motives always have elements of domestic politics because public funds will be used. Common practices include a gift of surplus food or the delivery of equipment that will provide additional employment in the donor country. Development aid may be supported as an international obligation, but only trivially if is perceived to be largely ineffective by western standards. Greater support may be obtained if the aid is linked to economic or international political interests of the donor.[10]

Every donor and recipient will reach an agreement that is in the self-interest of each. The self-interest of the donor in a development program is critical with respect to its particularity and its continuity. Plans are usually fragile and need modification on the basis of actual experience. The self-interest of the recipient is critical with respect to its political will and ability to do its job most effectively. Programs require continuing organizational development on the part of the host government and time, running into decades, for major progress to be made.

A common arrangement is a bilateral agreement for a project financed by a developed country in a less developed country. The countries will have exchanged ambassadors and will have many diverse issues under discussion, such as communications, monetary exchange rates, trade, and military assistance. Some issues may be political such as a voting record at the United Nations. The donor country will likely consider a development project as just one of the issues of the moment and bargain for advantages in other sectors in return for the project. The recipient country, too, may consider the project start as just an issue of the moment. It may also consider the project, following the start, to be its own business as a sovereign nation. All too frequently the financial controls over donated funds are less rigorous than over local funds, and donated supplies or equipment are stolen. In addition, it is likely to be discovered later that the project needs a change of plans or additional financing to be successful. All of these considerations tend to result in bilateral assistance projects of short duration and limited follow-up, since admission of difficulties could be troublesome to both parties.

Charitable and other private organizations have only a modest financial capability, and their self-interest is in providing useful assistance to groups of countries while maintaining a nonprofit or religious status. They do not need to be greatly concerned with the politics of host countries because they ask for little except a chance to work. They can be especially useful in working on long-term social problems common to several countries.

One long-term need is applied research of the kind that has been provided by the International Agricultural Research Centers.(11) The oldest one was set up in 1943 by the Rockefeller Foundation and the Mexican Government to improve the varieties of wheat and maize. The International Rice Research Institute began operations in 1962 in the Philippines. Numerous others followed, and all are banded together under the Consultative Group on International Agricultural Research (CGIAR) that was organized in 1971. The CGIAR provided a model for a fishery organization, the International Center for Living Aquatic Resources Management (ICLARM), that began operations in the Philippines in 1977.

Other charitable and religious organizations may assist with development through training and advisory programs concerned with such things as village water supplies, formation of cooperatives, and farm practices.

Private banks of the developed countries have provided large amounts of development assistance through their international operations by making loans to countries or to large industries. Their self-interest is to earn money on their depositor's money. Usually they operate as consortia which may extend credits to countries in the tens of billions of dollars.

Numerous defaults by less developed countries indicate that those countries could not or did not use the money as planned, nor could the banks that made the loans to them have been deeply involved in detailed project planning or follow up within the borrowing countries.

The World Bank, headquartered in Washington, D.C., and the International Development Banks of Asia, Africa, and Latin America have a different self-interest than the private banks.[12] These special development banks have money from governments which is to be used for various types of loans, including conventional loans at moderate rates of interest to governments and industries, as well as low-rate concessionary loans. These banks require performance in terms of project success as well as loan repayment. They are frequently involved in projects cofinanced by other donors. They also play a long continuing role by doing regular performance audits throughout the life of projects and a rigorous ex postevaluation. Such audits frequently lead to revised plans, including refinancing and reorganization of project personnel, as well as to more skillful planning of future projects.

A major multilateral effort is the United Nations Development Programmme (UNDP). This was established in 1965, through a combination of two earlier programs, with a broad mandate to assist with most aspects of development. It operates in collaboration with governments and other United Nations agencies which carry out its projects. These include feasibility studies, education and training, technical advice, economic and social planning, and investment stimulation. The locale of a project may be national, regional, or even global.

The UNDP obtains voluntary contributions, mostly from developed countries, and supports projects for which it normally provides about 60% of the cost; the remainder comes from the recipient country. About 80% of the project expenditures are made in countries with less than an annual per-capita income of US$500 equivalent.

A large portion of the UNDP projects is in its category of Agriculture, Forestry, and Fisheries. These, which were mostly multiyear projects, numbered 900 on September 30, 1982, and were budgeted at US$1430 million.[13] Of this total, 122 were fisheries projects which were budgeted at US$122 million.

2.6 LESSONS FROM EXPERIENCE

A wide diversity of scholarly views has been typical of early efforts that have failed their initial expectations. Much of the dispute continues among scholars who try to focus on narrow program strategy. Four emphases in programs since 1950 have been called fads,[14] each of which may still receive political support.

The first fad consisted of efforts to overcome oppressive governments through the introduction of constitutions and democratic organizations based on governments of Western Europe and North America. But such introductions failed to recognize the long social learning process and the need for educated and informed electorates that support the older democratic governments. The consequence was either no change in the ruling elite or an illusion of change to unrealistic alternatives.

The second fad was education and training. Primary education in rural areas induced people to migrate to city slums in search of better opportunities. Education at high school and university levels frequently produced people who could not fully use the skills they acquired because of lack of employment opportunities.

The third fad was to simulate the industrial complexes of developed nations with the expectation that the benefits would trickle down through increased employment and economic activity. This tended to induce further migration to city slums and neglect of agriculture.

The fourth fad was transfer of large amounts of money which could be used to provide infrastructure and support industrialization without recognition of the social and organizational changes necessary in government and business.

None of these fads was wrong by itself, but each was inadequate to meet the complex needs and failed in its expectations. Each was an attempt to find a trigger that would touch off a growth cycle. Is there a trigger? Mellor[15] and other development experts now suggest a rural-led, employment-oriented strategy in which the first priority is acceleration of agriculture, the second is expansion of small scale industry, and the third is expansion of exports to pay for necessary imports. All of these would require elements of each of the fads and, in addition, would require that effective government reach the local level. Kahn [4, p. 438] supports the importance of agriculture through his observation that few countries have been able to develop rapidly before their agriculture was in good condition. Donaldson[16] also argues that improvement of the private, family-run farms is critically important for economic growth in almost all countries.

If improvement of rural life and agriculture is a trigger, then surely the mechanism for action is a change in the image and function of rural community organizations. All too frequently, new national governments tend to centralize decisions to the exclusion of local governments, thereby leaving the most visible government functions in many societies as policing and tax collection--both of which are usually regarded as adversarial. Other functions are, of course, communications, education, and infrastructure of various kinds. But to these must be added the function of a trusted partner in development; a partner that knows the local people and their problems, that provides sound advice, and that sees promptly and honestly to the things that government must do to help, not dominate.

Traditional communities, typical of rural farming and fishing, are commonly less easily changed than urban communities where individuals find more diverse opportunities. And frequently they are affected differently by development programs than the urban communities. They represent, therefore, a special challenge for development planners.

NOTES AND REFERENCES

1. K. K. S. Dadzie, Economic development. *Sc. Am. 243* (3):59-65 (1980). Most of this issue was devoted to economic development.
2. (ICIDI) Independent Commission on International Development Issues (Brandt Commission), "North- South: A Programme for Survival." MIT Press, Cambridge Massachusetts, 304 pp., 1980.
3. D. Morawetz, "Twenty-Five Years of Economic Development 1950 1975." The World Bank, Washington D. C., 125 p., 1977. An excellent summary with many tables of basic statistics and a long bibliography.
4. H. Kahn, "World Economic Development: 1979 and Beyond." Morrow Quill, New York, 519 pp., 1979. See p. 64 ff.

5. M. Yudelman, The World Bank and agricultural development--an insider's view. World Resources Inst., *Paper* (1):1-36, (1985).
6. W. G. Zeylstra, "Aid or Development: The Relevance of Development Aid to Problems of Developing Countries." 2nd Ed., Sijthoff, Leyden, 269 pp., 1977.
7. F. H. Sanderson, The great food fumble. *Science* (Washington, D. C.), *188*:503-509 (1975). This issue was devoted to food problems.
8. FAO, *Food Outlook*, 1983 Statistical Supplement. 34 pp., FAO, Rome. March, 1984.
9. World Food Council, "The World Food and Hunger Problem: Changing Perspectives and Possibilities, 1974-1984. An Independent Assessment Presented to the World Food Council. 10th Ministerial Session, Addis Ababa, Ethopia. 11-15 June, 1984" 32 pp., 1984.
10. R. Cassen, R. Jolly, J. Sewell, and R. Wood, eds., "Rich Country Interests and Third World Development." Croom Helm, London, 369 pp., 1982. See especially Chapter I, Overview.
11. D. L. Plucknett and N. J. H. Smith, Agricultural research and third world food production. *Science* (Washington, D. C.), *217*:215-219 (1982).
12. K. Marsden and A. Roe, The political economy of foreign aid: A World Bank perspective. *Labour and Society* *8*(1):3-12 (1983).
13. United Nations Development Programme (UNDP), "Compendium of Approved Projects as of September 30, 1982." UNDP, New York, 1982.
14. E. Owens and R. Shaw, "Development Reconsidered." Heath, Indianapolis, Indiana, 190 pp., 1974. See p. 150 ff.
15. J. W. Mellor, "The New Economics of Growth: A Strategy for India and the Developing World." Cornell Univ. Press, Ithaca, New York, 335 pp., 1976. See Chapter XI.
16. G. Donaldson, Incentives and the farmer. *Finance & Development*, World Bank, pp. 17-20 (Sept. 1981).

Chapter 3

Fishery Systems

Fish to eat, fish to sell, fish to play with. This trilogy had its roots early in human history when communities were located where fishing for food was good, when people bartered fish or paid taxes with them, and when leaders sought the sport of fishing. The trilogy exists today with different social values in every society; values that persistently defy balancing by the best of governments.

Fish became scarce in some fresh waters centuries ago, but the fish in the oceans were regarded as inexhaustible until about the end of the nineteenth century. Then, with the advent of powered vessels, signs of depletion began on grounds close to ports. With the application of more power and ingenuity, signs of depletion have now appeared in almost all of the world's fisheries.

Fishing is regarded as a basic human right by most people who want to fish. Yet its antiquity, universality, and simple concept now mask a troublesome complexity. The acts of people catching fish result in fisheries--systems of activities that require supporting services, generate a product in the form of food or recreation, and stimulate government intervention because of too few fish in the commons to satisfy everyone.

It will be argued in the next chapter that the world's fisheries face a great transition, but before discussing the transition, it is desirable to describe briefly the characteristics of the fisheries at the end of the present era[1] with emphasis on the principal factors that must be considered by those who plan to develop them further.

3.1 TYPES OF FISHERIES

The development potential varies greatly according to the type of fishery. Unfortunately, no simple classification of the fisheries is practical, but the following four sets of categories have been used and provide points of departure.

Capture, Culture. Fishing and fish farming are the most clearly separated categories of the fisheries. The capture fishing seeks the resources of the commons, open to all who are able to participate unless restricted by governments. The fishers operate in all of the accessible public waters of the world where they compete with each other for the fish.

The fish farmers own their fish and own or lease the facilities in which they grow them. They usually control their production operations as all farmers do but, in some cases, must rely on the commons for brood stock, seed, or licensed growing areas.

Subsistence, Commercial, Recreational. Subsistence fishers pursue ancient practices of fishing in nearby waters for family food and for barter or sale in nearby villages. They work as individuals or in small groups with simple gear. They use few, if any, powered vessels or equipment and have little capital investment. Most of them live in rural areas where many are part-time farmers as well as fishers. Family members frequently assist with processing and sale of the catch. They are often very poor and, in some countries, regarded as people with an inferior occupation.

Commercial fishing is a business, and most commercial fishers depend on other businesses for supplies, vessels, equipment, repairs, and purchase of their catch. The scale of commercial fishing systems ranges from one-man motor vessels that supply fish to a local buyer for nearby markets to transnational corporations that operate fleets of vessels and prepare products for sale in national and international markets.

Another commercial fishery activity closely linked to both fishing and fish farming is the ornamental fish business. Colorful fish from both fresh and salt water are captured or farmed in warmer climates to supply a large international trade.

Recreational fishers are those who cherish and use the common right to fish, usually with primitive and inefficient methods. They value the experience of being close to nature, of traveling, of relaxation, and of companionship. Their numbers in the developed countries are growing rapidly and may be hundreds of times as great as the numbers of commercial fishers. Because of their numbers, they obtain political allocation of resources for recreational use and generate large business activities associated with their equipment, travel, and necessary services.

Small-Scale, Large-Scale. This distinction is usually applied to the capture fisheries and the term small-scale equated to artisanal which may include both subsistence and poorly equipped commercial fishers.

The small-scale fisheries are frequently conducted by poor people who obviously need assistance. Most development programs are designed for them and around more efficient fishing on the resources available to them.

The large-scale fisheries are those that already have substantial skills and financial resources to compete both for the resources and in national or international markets. But as the resources of the commons become fully utilized, they too may require development assistance.

Market, Market-Exempt. These categories of the capture fisheries separate, of course, the commercial and recreational fishers, but the market fisheries also include the subsistence fishers who depend on the fisheries for food, barter, or sale. The distinction is useful because all market-oriented capture fisheries have unique problems in which the businesses compete with each other for *both* the resources and the markets. When the resources will sustain no further increase in production these fisheries, either small- or large-scale, suffer severe economic problems. On the other hand, the market-exempt or personal fisheries for recreation can be sustained by allocation of the permissable catches among all people who want to fish--even to the point of requiring that catches be released alive. Such fisheries can be a much more valuable use of the resource than commercial fisheries.

3.2 RESOURCES

Each fisher depends on at least one but seldom more than about a dozen species of animals or plants. Yet the total number of species used is immense. The FAO in 1982,[2] listed more than 750 taxonomic categories of animals and plants (some containing more than one species), but only 20 species provide more than one-third of the world's fish production. The total is comprised predominantly of fishes but also includes crustaceans, mollusks, and a few species of numerous other groups such as marine mammals, turtles, sponges, and echinoderms. The plants are predominantly the several groups of large algae called seaweeds. Even so, this listing does not include the hundreds of additional species of animals and plants used for ornamental purposes.

Every species, indeed every race of a widely distributed species, has distinctive habits. These include all biological aspects of their lives, but one habit of special importance to the fisheries is migration. Migrations of a few oceanic species, such as some whales and tunas, may be transoceanic. But migrations of most resource species of fish in large fresh or coastal waters are a few dozen to a few hundred kilometers, depending on their usage of feeding and breeding grounds. Others, such as many crustaceans and shelled mollusks, may drift as larvae and then migrate little (if at all) during the rest of their lives.

The fishery managers, who are faced with this diversity of species, races, and migratory behavior, designate stocks which are units of the fishery resource that can be managed separately. Ideally, a stock is a single interbreeding unit of a race or species, but practically it is the one or more species caught by a single type of fishing operation in one locality.

Some resource organisms can be found in any of the waters of the world that are both accessible by fishers and suitable for the organisms. In general, such waters include the upper 500 m of the oceans in all parts not permanently covered by ice, as well as all of the fresh and coastal waters in which water quality is suitable for them.

But water is also a resource of the commons and a major factor in economic development. The fresh waters are consumed by cities, farms, and industry and used in place for power generation, transportation, and waste dilution. The coastal waters are used for ports, transportation, and waste dilution while the adjacent shores become cities and industrial sites.

All human uses of water change the environment of the fishery organisms for better or worse. For example, dams and reservoirs block migrations of fish in the rivers, decrease the area for riverine species, and increase the area for lake species. A little domestic sewage or animal waste fertilizes the water and enhances fish production, but too much causes eutrophication and may eliminate the dissolved oxygen essential for animal life. Water returned to a stream after use for industrial cooling may inhibit cold-water species and encourage warm-water species as long as the temperature remains tolerable for them.

All of these differences in the fishery resources and their environment mean that each fishery is unique. If the fishery is small-scale, it will harvest species in nearby waters, but the species may live throughout their lives in the same waters or merely migrate seasonally through those waters. If the fishery is large-scale, the vessels may be able to follow the fish for hundreds or even thousands of kilometers. Each stock will be fluctuating in abundance according to the amount of fishing throughout its range, and the natural or man-made changes in its environment.

Such migratory habits pose political problems because the habitat of a stock in larger lakes or rivers and in the ocean seldom coincides with any political boundaries. Only a few stocks in coastal waters of smaller countries will be completely within a single national jurisdiction. Even those stocks will probably span different local jurisdictions, each with different fishing practices and water usages.

3.3 FISHING METHODS

Each kind of fish in each kind of environment is caught most easily by gear adapted to both. The major kinds of gear now in use range from devices operable by one person without power to nets operable from a 4000-ton vessel which is equipped to catch, process, and preserve perhaps 100 tons of fish each day (table 3.1).[3]

Most of the extraordinary increase in world catch during the past three decades has been possible because of the technical innovations in fishing gear and associated vessel equipment. The use of power for propulsion and gear handling, which started toward the end of the nineteenth century, has continued to expand. Most fisheries have added larger vessels with longer range and better fish-finding and navigational equipment, and the largest have developed the ability to support fleet operations in any part of the oceans.

Further increases in efficiency have resulted from use of artificial fibers, especially nylon. Even the small-scale fisheries have been supplied with nylon gill nets, which are much more efficient than those made from natural fibers. And the large-scale fisheries have adopted immense nylon trawls or purse seines, which can be made stronger and larger than previous nets and which do not rot when stowed wet on deck while covered with fish slime.

The result has been an extraordinary increase in the range of catches per person. The approximate levels of catch in tons per man per year for representative kinds of fishing are shown in Table 3.2. The change in rate from the commonly accepted level of 1 ton/man/year for subsistence fishers without power increases in logarithmic steps with additional power, larger vessels, and better gear. The improvement possible with modern trawl-

Table 3.1
A Classification of Fishing Gear

1. Hand diggers and collectors--Rakes, tongs, or shovels for mollusks, crustaceans, or burrowing animals; knives and other tools used by divers to collect mollusks or sponges

2. Trained animals or birds

3. Spears and harpoons--hand-held spear with barb, arrow on a line, harpoon with detachable head for large fish, explosive harpoon for whales

4. Stupefying devices--poisons, explosives, electricity

5. Attracting devices--hook and line with lures or bait, broadcast bait, lights, simulated habitats

6. Herding devices--bubble curtains, lines, splashers

7. Stationary entangling nets--gill nets with single wall of mesh, trammel nets with multiwalled mesh

8. Stationary enclosures--pots for eels, crustaceans, octopi; brush or rock enclosures; small fyke or bag nets for river fish; large corral or trap nets for coastal fish

9. Mobile nets--lift nets (frequently used with a light), falling or cast nets, seines pulled on bottom toward a fixed point, floating purse seines, trawls pulled along bottom or in mid-depths

10. Towed dredges--rake-like bar fitted with a bag for oysters, scallops; conveyor-type assisted by hydraulic jets for clams

Source: W. F. Royce, "Introduction to the Practice of Fishery Science." Academic Press, Orlando, 1984.

Table 3.2
General Production Levels of Various Fishing Practices

Catch (tons/man/year)	Fishing Method
0.01	Occasional recreational angling
0.03	Average recreational catch in the United States
0.1	Regular recreational angling
0.3	Seasonal subsistence fishing with lines, traps, spears, or nets.
1	Subsistence fisheries from manually operated boats
3	Subsistence or primitive commercial fisheries from small power boats
10	Commercial fishing from small coastal vessels for highly valuable fish
30	Commercial fishing from medium to large vessels for moderately to highly valuable fish
100	Commercial fishing from best modern trawlers and purse seiners for moderately valuable fish
300	Commercial fishing from purse seiners for low value fish such as anchovies, pilchards, menhaden, or jack mackerel
1000	Commercial fishing from the most efficient modern purse seiners or trawlers for low value fish when abundant and when deliveries can be made daily

Source: adapted from W. F. Royce, "Introduction to the Practice of Fishery Science." Academic Press, Orlando, 1984.

ers and purse seiners fishing for species similar to those taken by many subsistence fisheries is a hundredfold; for lower value schooling species, it may be a thousandfold. In contrast recreational fishing is popular at catch levels between about one-tenth and one-hundredth the level of subsistence fishing and less than one-thousandth the level of most commercial fishing.

3.4 FARMING METHODS

When the stocks of wild plants and animals can not supply the demand, farmers domesticate a few desirable and amenable species and then control their lives in order to increase the production rates. The harvest rates of wild fish from natural waters can range from about 10 kg/ha/year in the most barren waters to about 200 kg/ha/year in the most pro- ductive. Much greater rates can be attained by fish farmers.

The most widespread method of fish farming is pond culture, in which at least part of the animal's food is supplied by managing the pond. Many of the animals are omnivorous, eating mixtures of natural plant and animal food. These include the several Asian carps, catfish, goldfish, milkfish of southeast Asia, bait fishes of several species, shrimps, prawns, and crayfish. Some of the animals require fresh water, in which case the ponds may be constructed and operated in combination with irrigation systems for field crops. In these situations the ponds are often fertilized or the fish fed by wastes from field crops or other farm animal production. Other animals such as shrimps and milkfish are grown in brackish coastal waters. Sometimes, combinations of species that use different parts of the ponds can be grown efficiently. Farm pond production rates can regularly be maintained at levels of 2000 kg/ha/year but the maximum under precisely controlled conditions appears to be about 10,000 kg/ha/year.

The second major fish farming system is raceway or cage production of animals that are fed a complete diet. These are carnivorous animals such as salmons, trouts, basses, porgies, pompanos, pikes, eels, and yellowtail. They are grown in either fresh or salt water according to the animal's tolerance. This system permits much greater production rates--even 500,000 kg/ha/year--but it is practical only where suitable water and low-cost food are available. It requires flowing high quality water at temperatures nearly ideal for the particular animals and usually containing at least 4 ppm of dissolved oxygen. Since most of the animals are carnivorous, their diet must be nutritionally balanced with a protein content usually between 20 and 35%.

The third major farming system is salmonid ranching. The salmons, and a few other anadromous species, can be propagated in fresh waters, released to feed in the sea, and captured upon return to the home stream. The total production from such operations exceeded 150,000 tons in 1985 and was growing rapidly.

The fourth major farming system is for the filter feeding mollusks: oysters, clams, mussels, and scallops. These are grown in clean, brackish, coastal waters where the tidal circulation brings food to them. Such free food is an obvious advantage, but the animals cannot be isolated from competitors or predators, and the difficulties of maintaining sanitary growing conditions are rapidly increasing.

The fifth farming system is for seaweeds, the large leafy marine algae, of which several species are grown in eastern Asia. These are grown in coastal waters by controlling the reproductive cycle and providing rocky bottoms or rafts for attachment of the plants.

A sixth and widespread small farm system is for ornamental plants and animals. Hundreds of species are grown, each according to its needs in small aquaria or tanks. Most of the species come from tropical countries and supply large markets in the developed countries.

Fish farming is rapidly becoming much more sophisticated. More (but not enough) scientific attention is being given by governments and businesses throughout the world to domestication of promising wild species, selective breeding, disease control, better diets, superior products, international marketing, and pond engineering. The farming systems are beginning to include seed (egg or larval) production facilities, premixed diets distributed by feed companies, international transfer of seed or brood stock, and other practices paralleling the trends in agriculture.

3.5 RECENT TRENDS IN OVERALL PRODUCTION

The total fishery production of the world in the early 1980s was about 90 million tons round weight. No complete record is available but FAO, which compiles the best available statistics, reported 74,517,900 tons of commercial fishery production in 1983.(2) This figure does not include marine mammals, aquatic plants, ornamental fishes, recreational catches, and a large part of fish farm production which can only be roughly estimated to make up the balance.

Trends in the commercial fish production since the late 1940s are shown by the annual FAO data (Fig. 3.1). From about 1950 to 1970, there was an annual rate of increase in the total world production of about 6%, but since then it has slowed to about 2%. The production by continents has varied. South American production grew very rapidly until about 1970, but then declined due to the collapse of the Peruvian anchovy fishery. Asian, African, and U.S.S.R. production more or less followed (or set) the worldwide trend, but European and North American production grew much more slowly during the 1950s and 1960s. Then a new pattern emerged in the late 1970s. European, U.S.S.R., and African production nearly leveled out or declined. Asian production increased again and South American production regained part of its earlier status with development of new fisheries on stocks off Chile. North American production grew more rapidly, apparently because of the exclusion of some foreign fishing and the increases in domestic production of Canada and the United States. Thus the modest in-

Fig. 3.1. Trends in world production from market fishing. (Points are averages of 5 or 3 years.) Source FAO, "Yearbooks of Fishery Statistics: Catches and Landings, 1948-1983".(2)

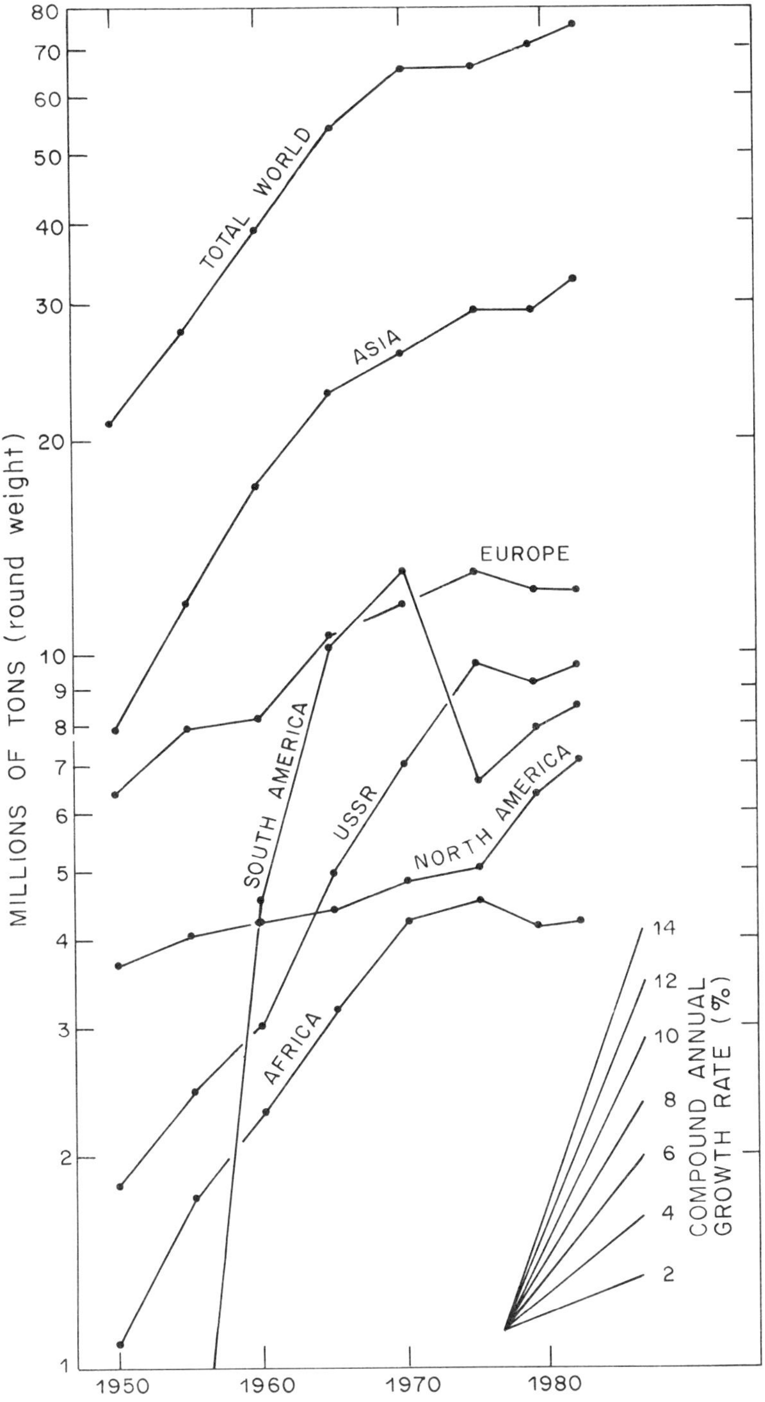
MILLIONS OF TONS (round weight)
80
70
60
50
40
30
20
10
9
8
7
6
5
4
3
2
1
1950
1960
1970
1980
TOTAL WORLD
ASIA
EUROPE
SOUTH AMERICA
USSR
NORTH AMERICA
AFRICA
COMPOUND ANNUAL GROWTH RATE (%)
14
12
10
8
6
4
2

crease in world production after the middle 1970s was due to the increases in Asian, North American, and South American production.

Fish farm production statistics are much less complete, but the best estimates from FAO (excluding aquatic plants and ornamental fish)[4] show world totals of 4,620,000 tons in 1975 and 6,582,000 tons in 1980. The 1980 total included about 1% crustaceans, with the balance nearly equally divided between mollusks and finfish. The geographic distribution of the 1980 production was about 79% from Asia, 17% from Europe, and 3% from Canada and the United States (See Section 4.2.2).

Recreational fishery catch statistics are available from few countries, but the social and economic dimensions of the recreational fisheries are very large in Canada, the United States, and the countries of Europe. In most of these countries these fisheries are the dominant (if not the only) significant fisheries in the inland waters, and they are important in the coastal waters.[5] In the United States, which has more complete statistics than most others, between 55 million and 60 million anglers participate in fresh and saltwater fishing and the economic benefit of their direct expenditures in 1980 was estimated to total US$17.3 billion, dwarfing the economic value of the commercial catches.[6]

3.6 PROCESSING AND PRODUCTS

Fish are highly valuable in human diets. The edible part of their bodies is typically about 17% protein (wet weight basis), and from about 1 to 20% fat, depending on the species and season. The fish protein from most species is as good for human consumption as mammalian or avian proteins, whereas the fish oils are generally superior because of a higher proportion of polyunsaturated fatty acids. And the choice of oily or nonoily species provides a diversity of flavor and caloric content. Fish proteins are also valuable in domestic animal diets. About 30% of current world production is caught for reduction to fish meal used in animal diets and, in addition, much of the waste from fish processing plants is also prepared for animal diets.

The fisheries also produce other valuable items. In addition to ornamental fish, there are the bait fish used by recreational fishers, the fish oils used in paints and margarine, the colloids from algae used as thickening agents, the special leathers, and the ornaments from mollusks and corals.

Fish is also one of the most difficult kinds of human foods to preserve in good quality for human consumption. The capture fisheries commonly catch the animals after a struggle, in quantities that cause many to be bruised, under conditions that inhibit immediate evisceration and bleeding, and days before ultimate processing. These conditions, of course, vary according to the kind of fishery and the kind of fish, but rarely can the conditions of capture approach the sanitary standards of a modern poultry or meat processing plant.

After capture, the simplest method of all is distribution and consumption of the fish within the day or two that they will retain acceptable quality without chilling, or within the one to three weeks that they can be kept if thoroughly chilled. This method was used for nearly half of the world production in 1950 but, in spite of the increasing availability of ice, the proportion declined to about 20% in 1982.[2]

The traditional curing methods of drying, salting, smoking, pickling, and fermenting, which have evolved over the centuries, persist in many forms. They usually require little capital investment and are suitable for many species and products. In the developing countries, some of these methods are used for most of the low cost fish that cannot be consumed fresh. But in or for sale to the developed countries, refined methods may produce choice products such as caviar or smoked salmon. The use of these methods, like consumption while fresh, has not kept pace with the increase in catch--the proportion cured declined from about 25% in 1950 to about 15% in the early 1980s.

The capital-intensive methods of fish processing have been expanding more rapidly than overall catch. Freezing was usually a method that produced less desirable products in the 1940s because it was used to "save" fresh fish from spoilage. But techniques have been greatly improved and the proportion of fish frozen in the world increased from about 5% in 1950 to about 22% in the early 1980s.

Canned fish production, which had been a major method of long-term preservation in the first half of the twentieth century, involved less than 10% of world catches during the 1950s and 1960s but by the early 1980s had increased to about 14%.

Fish meal and oil production has been the fastest growing kind of processing. Less than 10% of the catches were taken especially for fish meal in 1950, but this rose to 38% in 1970. Later, the proportion declined with the collapse of the Peruvian anchovy fishery but still involved about 30% of world production by the early 1980s. And, in addition to the fish caught especially for fish meal and oil, much of the waste from freezing and canning plants in recent years has become a profitable by-product, although no good statistics are available.

With the expanding importance of fish products for human and animal food, the issue of product quality has received much more attention. Inspection systems have been developed by industry-wide and government efforts. Quality standards and codes of practice have been formulated in many countries, in part by international collaboration through the Codex Alimentarius Commission and the International Commission on Microbiological Specifications for Foods.

3.7 MARKETS

The tripling of world fish production and the major trends toward large-scale capital-intensive processing methods have been accompanied by major changes in the marketing concepts. These changes have been possible largely because of improved quality.

Distribution fresh or cured, that comprised about three-fourths of total production in 1950, declined to about 35% in the early 1980s. This requires marketing of many species and products of uncertain quality. Such products are seldom standardized, indeed may vary in quality from day to day. They are handled typically by small-scale wholesalers and retailers in small lots in nearby markets. The consumer is likely to judge quality by personal inspection and personal acquaintance with the retailer who may similarly judge quality by his inspection and acquaintance with the wholesaler. Distribution of these products did increase in quantity from 1950 to the early 1980s, but only from about 15 million to 26 million tons round weight--an increase that is much less than the population increase in most nearby markets.

The processing and packaging of frozen fish was greatly improved during the 1950s. Fish of better quality were frozen and packaged to be convenient for homes and food service institutions. This led to increases in dealer and consumer confidence in the products. This, in turn, led to handling in larger lots and direct buying in quantity from processors by large retailers or food service institutions. The specialized fish dealers were no longer necessary to handle the small lots of variable quality. The improvement in quality has continued with increasing fractions of the catch being processed and frozen under excellent conditions within hours after capture. The total quantity of frozen fish increased from about 1 million tons in 1950 to about 16 million tons (round weight) in the early 1980's, and continues to ncrease rapidly from year to year.

Recently included in the frozen products are very high quality reformed fish made by deboning and mincing the flesh. One product, rapidly increasing in sales, is surimi--a white, high protein food made from nonoily fish that can be flavored to imitate other species.

Canned fish, which have been of more assured quality for several decades, have also increased in quantity. About 2 million tons were canned in 1950, about 10 million tons (round weight) in the early 1980s.

Fish meal and oil have been produced in large volumes at relatively few locations and most is handled in large lots by feed companies. About 2 million tons were caught for fish meal in 1950 increasing to about 22 million tons in the early 1980s.

Perhaps the best indicator of the long distance marketing of fish products is the proportion in international trade. It increased from about 20% of catches in 1950 to about 33% in the early 1980s, an increase in volume of about 6 times. A substantial proportion of the recent trade is fish meal, but much of the rest is relatively high priced frozen and canned products, such as shrimp, a large majority of which is sold internationally.

The general pattern of international fish trade follows the world food trade in the sense that most of it is among developed countries, and when developing countries are involved they export about three times as much as they import. The net flow of fish is away from the hungry but their exports are predominantly high-priced and their imports predominantly low-priced.[7]

3.8 GOVERNMENTS

Governments have seldom been considered previously as a major part of fishery systems, but their role is becoming critical with respect to all resources of the commons as those resources become scarce. Most of the world's wild fish and water resources are now so scarce that governments play major roles in their use.

Some intervention is not recent; ancient communities occasionally exercised some control over fishing. And, if government is considered in its broadest sense to include traditional community customs, such as understandings about where fishers may operate their gear, or the belief in a public right to go fishing without interfering with another's right, limited intervention has been widespread. But a surge in national government intervention began early in the twentieth century after the great public concern about the destruction of natural resources during the nineteenth century.

Unfortunately, the recent government policy for the fisheries involving custom, law, and bureaucracy has not been widely documented for many countries. It involves diverse functions of every national government with various aspects performed by regional or local governments and by local custom. In addition, some governments have delegated a few functions to international organizations.

Although the details of policy are diverse, many aspects of fishery management and development in developed countries are recorded. These have provided models for many other countries, although the relative importance of each aspect will be different in every situation.

But fishery policies are only part of the complex area of public policymaking with respect to scarce natural resources.(8) Policy analysis of the entire area is difficult, embodying as it does the realm of social and political philosophy. It is especially difficult when dealing with the basic resources of water, land, and food. It is even more complex when dealing with resources that must be rationed by government instead of the marketplace. The analysis must deal with people's perceptions of values and political solutions as well as technical facts and administrative solutions.

Fishery policy has received little scholarly attention from political scientists. The following description of existing fishery policies is, therefore, largely a personal perception based on a general approach to policymaking.(9) It is also a prelude to discussion in subsequent chapters of necessary policy evolution.

3.8.1 Fishery Management Policies

Fishery management policies in most developed countries now include a principal goal of conservation of the fishery resources and a secondary goal of using them for economic development. But the policies and activities that contribute to the two goals are mostly very different, and the relative emphasis on each goal varies greatly among localities. Therefore, discussion of fishery management policies will emphasize those required for conservation, and discussion of fishery development policies in the next section will emphasize those required for increasing social benefits, even though there is some overlap in activities.

Public Perceptions. The widespread public concept of fishery management is conservation, meaning wise use, preservation from loss or waste, and official supervision of a resource of the commons that can be used by everyone. The concept has evolved from dependence on fishery resources by many coastal peoples, and acceptance of the reality that the fishery resources could not be conserved if everyone used them freely.

The concept of a universal right of free access to fish resources is ancient (See Chapter 5). The right has been firmly retained by most peoples but has been modified by acceptance of official restrictions on efficiency when those are based on research rather than merely on a political process.

Fishery Agencies. Most people agreed long ago that the fishery issues required sound technical information on the resources and the environmental quality that they require plus expert action to implement policies. They have, therefore, supported specialized fishery agencies in governments with authority for numerous activities.

But many other agencies must consider fishery problems. They include the police, which may enforce fishery regulations; water management agencies, which may change the aquatic environment; and ministries of foreign affairs, which must consider problems of stocks shared with neighboring countries. The agencies may include those with local, regional, national, and international authorities, all of which must cooperate on the fishery issues.

Research and Statistics. Soon after fish conservation became a public issue in the late nineteenth century, biological research began on fish and fish stocks. It has steadily expanded to include most aspects of the fish themselves and their environment. Even earlier, statistics on the amount of fish caught were included in business publications; the statistics have been expanded to include details of the fishing effort, species, sizes, age, and location of catches that can contribute to knowledge of the biology of the species. Much of the recent research is directed at improvement of the catch statistics, understanding the relationship between the amount of fishing and the sustainable yield of each stock, and determining the impact of environmental change. The research is used to inform the public of the nature, condition, and use of the resources. This research is mostly applied and continuous, providing recurring recommendations, usually annually, for regulation of fishing.

Laws and Regulations. Both laws and regulations are commonly used. The laws are legislative actions to provide basic guidance and to delegate limited authority to fishery agencies. These agencies usually have authority to make regulations that modify the basic laws such as catch limits, closed seasons, closed areas, and gear restrictions.

Enforcement of Laws and Regulations. In many countries the fishery laws and regulations are complex and so different from most other laws that special enforcement officers are employed by the fishery agency. The enforcement activity is essential if public confidence is to be maintained. It is also expensive; frequently requiring a large proportion of agency budgets. Further, it may be ineffective if the need for the laws and regulations is not broadly accepted by the fishers.

Protection and Improvement of the Aquatic Environment. The rapidly increasing use and abuse of water changes the fish environment. Many fishery agencies have been given authority to review plans for use of water or the adjoining shorelands that may affect the fisheries. They may recommend general policies or water quality standards and approve or reject construction plans. In effect, because they have extensive information on the resources and the environment, the agencies become political advocates of the needs of the fisheries.

Protection of Public Health. Some aquatic foods from certain mollusks and fish, may be toxic, unwholesome at certain seasons, or easily contaminated during handling. The expertise of the fishery agency is relied upon to recommend regulations for protection of the public, usually in collaboration with public health agencies.

Propagation and Distribution of Fish. The demand for fish is so great and the condition of the waters so poor in many developed countries that fishery agencies enhance the natural populations by regularly stocking the native species or by introducing desirable species in new locations. Such practices are usually restricted to the nonmarket fisheries in which the value of each fish may be very high.

International Cooperation. Every national government has fish and water resources that cross international boundaries, and it cannot exercise effective authority over those resources without cooperating with neighbors. The fishery management for conservation objectives requires agreement on uniform statistics, joint research, and coherent policies.

Role of Business. Fishing, fish farming, fish processing, fish marketing, and support of recreational fishing are performed by the private sector in most countries. The businesses contribute to the economy by providing a service and employment. They perform many of the important organizational tasks of society, and the managers of businesses discharge public functions that matter to many people.

The managers or executives usually work closely with government executives because arbitrary government action can disrupt the business, and if large businesses fail, that can disrupt the government. Consequently, fishery management (along with many other government activities) functions best when it recognizes the needs of business.[10] Many countries, anxious to better manage or develop their fisheries, have tried to form parastatal corporations which have usually failed to survive economically. It is essential therefore, for the success of fishery management or development, that government and fish businesses have a good working relationship.

3.8.2 Fishery Development Policies

Fishery development policies are considered here to be those other than conservation that optimize the overall social benefits derived from the fisheries. Such benefits are primarily for the people engaged in the businesses of the fisheries on a commercial or subsistence basis or for the businesses in support of recreational fishing. Secondary benefits may also accrue to communities through increases in food supply or economic activity.

Successful fishery development policies are much more difficult to execute than fishery management policies. The goal of improved social benefits from the fisheries must be approached with programs tailored to the values held by different fishing communities, whereas the goal of conservation is commonly approached with national programs tailored to the different stocks of fish. Furthermore, the theory and practice required for conservation programs have been widely accepted since the 1950s, whereas the theory and practice of fishery development programs have yet to be proven.

Public Perceptions. The prevailing perception of fishery development is an action to increase the supply or availability of fish, thereby providing more employment and more earnings for the fishery community, and perhaps more export earnings or less food imports for the nation as a whole. This perception is reinforced when fishers are relatively poor and fish are a major food of their communities. A common result is acceptance of a large range of special government services to the fisheries in addition to management for conservation.

An additional perception of extraordinary importance for the market-oriented fisheries is freedom for everyone to fish. As will be discussed at length in subsequent chapters, when this freedom is exercised and leads to excess fishing effort, the customary reduction of efficiency by fishery regulation eventually causes economic disaster.

Fishery Agencies. The fishery management agencies usually handle the fishery development activities. But other agencies may be involved in specialized assistance, such as development banks or agencies, small business advisory agencies, and extension services to fishers and fish farmers. Other agencies may greatly influence fishery policy either favorably or adversely through price fixing, import controls, and vessel licensing. Still other local institutions may greatly influence the fisheries through infrastructure, such as ports, roads, cold stores, energy supplies, and specialized education.

Research and Statistics. Most of the government statistics on the fisheries are collected and distributed to assist in business decisions. These include catches by gear, species, and size that may also be important to fishery management, but many others may be provided, such as catch trends, price trends, daily market supplies and prices, quantities of commodities in storage, foreign trade, and employment.

Government research for development of the capture fisheries is also largely completely different from the research for management. Much of it is directed at the business decisions about processing methods, product quality, product development, and marketing. A different area involves improvement of vessels and fishing gear, especially for underutilized species.

The research assistance to fish farmers may also involve processing, products, and marketing, but the major areas are more likely to be related to farm production methods. Such areas are animal nutrition, selective breeding, pathology and disease control, water quality, and pond engineering.

Advisory Services. The best of these are extension services operated in close collaboration with research institutions. The essential feature is rapid delivery of research to its users and reciprocal transfer of research needs from users to institutions. Demonstration projects and training services may be included for both fishing and fish farming.

Infrastructure. Roads, energy supplies, water supplies, harbors, freezing plants, and other facilities may be essential for fishery development. The needs will vary with each location.

Fishing or Farming Equipment and Supplies. These may include vessels, fishing gear, fish seed, and a variety of supplies. When they are lacking, governments may decide to supply them instead of relying on private business. Such an arrangement may be desirable temporarily, but widespread experience with such government activities has shown that these functions are better performed by the private sector.

Low-Cost Loans. Small-scale operators in fish businesses may have difficulty obtaining credit, because small loans are expensive to handle and because the lending institutions have too little experience with the business. This may be overcome by providing government guarantees for loans and training for officers of lending institutions in handling the loans. Such loans may also be temporary subsidies extended to entire fisheries after resource failures due to environmental changes or permanent subsidies after persistent overcapitalization of fisheries on overfished resources.

Social Services. In a few countries, fishers are included with merchant seamen who are eligible for subsidized health and special educational services.

International Cooperation. This may start from the cooperation essential for conservation but will eventually include additional political factors. It may involve allocation of a portion of the harvest to foreign vessels, control of the foreign fishing through licensing and fees, agreement on joint enterprises, division of a shared stock with a neighbor, and decisions about the proportion of a catch to be exported.

NOTES AND REFERENCES

1. Part of this chapter is condensed from modern treatises on fishery management. Included are the following:
   J. A. Gulland, ed., "Fish Population Dynamics." Wiley, New York, 372 pp., 1977.
   W. F. Royce, "Introduction to the Practice of Fishery Science." Academic Press, Orlando, Florida, 428 pp., 1984.
   J-P. Troadec, Introduction to fisheries management: Advantages, difficulties and mechanisms. *FAO Fish. Tech. Pap.* (224):1-57 (1983).
2. FAO, "Yearbooks of Fishery Statistics." (The series on Catches and Landings began in 1948 and continues to the present. Beginning in 1963, the series on Catches and Landings was issued with even numbers and accompanied by the series on Fishery Commodities with odd numbers.)

3. A more detailed international classification of fishing gear is in use. See C. Nedelec, Definition and classification of fishing gear categories. *FAO Fish. Tech. Pap.* (222):1-51 (1982).
4. FAO, A study of methodologies for forecasting aquacultural development. *FAO Fish. Tech. Pap.*, (248):1-47 (1984).
5. J.-L. Gaudet, Introduction and summary, *In* J. H. Grover, ed., "Allocation of Fishery Resources: Proceedings of the Technical Consultation on Allocation of Fishery Resources, Vichy, France, 20-23 April 1980." FAO and Amer. Fish. Soc., pp. 1-7, 1982.
6. R. F. Hutton and J. H. Kutkuhn, Angling in the United States of America. *In* R. H. Stroud, ed., "World Angling Resources and Challenges: Proceedings of the First World Angling Conference, Cap d'Agde, France, September 12-18, 1984." Int. Game Fish Assoc., Fort Lauderdale, Florida, pp. 83-100, 1985.
7. G. Kent, The pattern of fish trade. *ICLARM Newsletter*, 6(2):12-13, (1983).
8. S. Welch and R. Miewald, eds., "Scarce Natural Resources: The Challenge to Public Policymaking." Sage Publications, Beverly Hills, California, 1983.
9. C. E. Lindblom, "The Policy-making Process." 2nd Ed. Prentice-Hall, Englewood Cliffs, New Jersey, 131 pp., 1980.
10. *Ibid.* See Chapter 9, The privileged position of business in policy-making, pp. 71-82.

Chapter 4

The Transition of the Fisheries

The seeds of change were sown in the 1970's but the fruit has not yet ripened. The expectations of bountiful harvest were formed early in that decade when, after 20 years of unprecedented growth in world fishery production, opportunities for much more growth and a redistribution of benefits seemed possible. The widespread perceptions were brought to a political focus by FAO in the early 1980s by its linkage of a "New International Economic Order" for developing countries to "their rightful place in world fisheries" (see Chapter 1).

Respected scientists had estimated potential annual ocean yields of animals of 200 million to even 2,000 million tons.[1] The Assistant Director General for Fisheries of FAO pointed out in a speech in 1970 that the fisheries contributed about 11% of the animal protein consumed by man and that the percentage should increase considerably in the future.[2]

Knowledge of the world's ocean fishing also became widely distributed through the publications of FAO. Its fishery statistical series that started in 1948 revealed that a large part of the fish catches during the 1960s and 1970s was made by a few fishing nations that were operating distant water fleets, and these were operating predominantly close to the coasts of other countries.

Those coastal countries gained control of their EEZs in the mid-1970s along with obligations for scientifically based fishery management. Such management has required formation of new national and international organizations that will impact all fishing businesses in the world.

Other uses and abuses of the waters had also become widespread by the 1970s. Damage to freshwater and inshore fisheries from environmental change was responsible in part for rapidly increasing concern about environmental quality. It resulted in international recognition that environmentally sensitive and responsible development was essential for long term economic development. A declaration of the United Nations Conference on the Human Environment in 1972 stated a conviction that "international organizations play a coordinated, efficient and dynamic role in the protection and improvement of the environment."(3)

If the decade of the 1970s was the decade of new hopes, the decade of the 1980s appears to be the decade of realization of how difficult solutions to the problems will be.

4.1 NEW LAWS

Unilateral claims of fishery authority beyond the customary three-mile zone spread rapidly during the 1960s after United Nations Conferences on the Law of the Sea (UNCLOS) in 1958 and 1960 failed to reach agreement on a wider zone. The third conference reached a consensus in 1974 on an EEZ 200 miles wide, and on many other aspects of fishery management in the coastal zones. These were debated further, and finally a Convention was adopted in 1982.(4)

The text of the proposed convention(5) (Articles 45 and 46) gave coastal states the right to claim an EEZ up to 200 miles wide within which it has "sovereign rights for the purpose of exploring and exploiting, conserving and managing the natural resources, whether living or nonliving...." This nationalizes about 95% of the well-explored stocks of fish in the oceans--essentially all except a few species that make transoceanic migrations. It happens because almost all of the good fishing grounds are within 200 miles of a coast, and the central ocean is relatively barren in low or middle latitudes of both hemispheres.

Within this zone (Article 50) "the coastal State shall determine the allowable catch of the living resources..." but with substantial obligations for their conservation. "The coastal State, taking into account the best scientific evidence available to it, shall ensure through proper conservation and management measures that the maintenance of the living resources in the exclusive economic zone is not endangered by over exploitation. As appropriate, the coastal State and competent international organizations, whether subregional, regional or global, shall cooperate to this end.

"Such measures shall also be designed to maintain or restore populations of harvested species at levels which can produce the maximum sustainable yield, as qualified by relevant environmental and economic factors, including the economic needs of coastal fishing communities and the special requirements of developing countries, and taking into account fishing patterns, the interdependence of stocks and any generally recommended subregional, regional or global minimum standards.

"In establishing such measures the coastal State shall take into consideration the effects on species associated with or dependent upon harvested species with a view to maintaining or restoring populations of such associated or dependent species above levels at which their reproduction may become seriously threatened.

"Available scientific information, catch and fishing effort statistics, and other data relevant to the conservation of fish stocks shall be contributed and exchanged on a regular basis through subregional, regional, or global organizations, whether, where appropriate, and with participation by all states concerned, including States whose nationals are allowed to fish in the exclusive economic zone."

Additional obligations for the utilization of the living resources included the following (Article 51): "The coastal State shall promote the objective of optimum utilization of the living resources in the exclusive economic zone....

"The coastal State shall determine its capacity to harvest the living resources of the exclusive economic zone. Where the coastal State does not have the capacity to harvest the entire allowable catch, it shall...give other States access to the surplus of the allowable catch.

"In granting access to other states...the coastal State shall take into account all relevant factors, including...the economy of the coastal State concerned and its other national interests...the requirements of developing countries...and the need to minimize economic dislocation in States whose nationals have habitually fished in the zone or which have made substantial efforts in research and identification of the stocks.

"Nationals of other States fishing in the exclusive economic zone shall comply with the conservation measures...and regulations of the coastal State. These... may relate...to the following:

(a) Licensing of fishermen, fishing vessels, and equipment, including payment of fees...;

(b) ... Species which may be caught...quotas of catch...during any period;

(c) Regulating seasons and areas of fishing, the types, sizes and amount of gear, and the numbers, sizes and types of fishing vessels that may be used;

(d) Fixing the age and size of fish...that may be caught;

(e) Specifying information required of fishing vessels, including catch and effort statistics and vessel position reports;

(f) Requiring...fisheries research programmes...and reporting of associated scientific data;

(g) The placing of observers or trainees on board such vessels....

(h) The landing of all or any part of the catch...in the coastal State;

(i) Terms and conditions relating to joint ventures or other cooperative arrangements;

(j) ...training personnel and the transfer of fisheries technology, including...fisheries research;

(k) Enforcement procedures."

Special provisions are made for species with life habits that cause unusual international problems. When stocks occur within the EEZs of two or more coastal states, "these States shall seek...to agree upon the measures necessary to...ensure the conservation and development of such stocks..." (Article 52). When stocks are highly migratory beyond the EEZs, the states whose nationals fish for them "...shall cooperate with a view to ensuring conservation...and optimum utilization...both within and beyond the exclusive economic zone..." (Article 53). When stocks are anadromous, the states in whose rivers they originate "...shall have the primary interest in and the responsibility for such stocks (Article 54).

The coastal states were also to seek agreement and to cooperate with neighboring states on broad aspects of conservation and utilization, and to prevent, reduce, and control pollution or dumping in oceanic or coastal areas.

After a consensus was reached on fishery aspects of international law in 1974, the provisions were widely accepted even though a convention had not been completed. Most coastal states, anticipating large benefits, soon adopted national legislation or other policies consistent with certain provisions, especially the extension of the EEZ, usually to 200 miles. Most of the developed nations with important fisheries extended their conservation and utilization policies and practices, i.e., their fishery management, to their EEZs during the late 1970s. Even so they have all experienced major difficulties with the effectiveness of their fishery management, especially because international agreement is still needed on management of most major stocks.[4]

The other new laws of major importance to the fisheries are for protection of the environment. Numerous countries have adopted laws for this purpose, most of which are more or less based on a national law of the United States, the National Environmental Policy Act of 1969 (NEPA). This act, in a declaration of policy, directed all agencies of the Federal Government to:

(A) utilize a systematic, interdisciplinary approach which will insure the integrated use of the natural and social sciences and the environmental design arts in planning and in decision making which may have an impact on man's environment;

(C) include in every recommendation or report on proposals for legislation and other major Federal actions significantly affecting the quality of the human environment, a detailed statement by the responsible official on--(i) the environmental impact of the proposed action, (ii) any adverse environmental effects which cannot be avoided should the proposal be implemented, (iii) alternatives to the proposed action, (iv) the relationship between local short-term uses of man's environment and the maintenance and the enhancement of long-term productivity, and (v) any reversible and irretrievable commitments of resources which would be involved in the proposed action should it be implemented.

The required detailed statement has become generally known as an Environmental Impact Assessment (EIA)[(3)] [or as an Environmental Impact Statement in the United States (EIS)], and the procedures have been somewhat standardized.[(6)] But the EIA is just an identification of the problems of environmental quality in a project, and the administrative mechanism for control of environmental insults has barely begun to be effective in only a few countries.

4.2 FUTURE SUPPLIES

Any estimates of fishery development potential require estimates of future production which must start with recent trends (Section 3.5). But the transition of the fisheries is still under way, so no simple projection of the trends can provide a satisfactory estimate of the future. The difficulties are indicated by the differences in projections made after the mid-1960s, and by the differences among recent authors which have arisen largely because of the complexity of the fisheries.

Various projections have been made by the FAO since the mid-1960s, and the discrepancies between the early projections and the actual events have illustrated the difficulties. Their projections through 1975 were a basis for the fishery part of the Global 2000 Report to the President (of the United States) in 1980, which was quite pessimistic about prospects for significant growth in world production.(7) A criticism of the Global 2000 Report and a more optimistic view has appeared,(8) but the best available projection is probably the most recent one prepared by M. A. Robinson of the FAO.(9) He points out that catches of conventional types of fish might possibly increase by 20 million to 30 million tons annually, but at least half of the increase could be achieved only by better management of, or favorable ocean conditions for, the depressed stocks. Increased fishing pressure appears likely to increase catches only from a few stocks in the southwest Atlantic.

4.2.1 Fishing

The global potential for fish catches during the next decade or two appears to be a rate of increase a little less than the recent rate of increase in population--about 1.8% annually. The total production from the marine fisheries will probably approach a ceiling of less than 30 million tons greater than the current production of about 70 million tons. But any overall rate masks complex problems which will vary immensely among stocks of fish and among countries.

The supply, rather than the market, will probably be the major limiting factor. Demand for fish as food will probably increase even more rapidly because of the higher rate of increase in population, the improvements in products, and the more widespread marketing. Average prices have been increasing and are expected to increase more rapidly than supply.

This projection of future catches is based predominantly on the marine animal resources that supply about 80% of the total. Freshwater, fish-farm, and aquatic plant production account for the balance (based largely on less reliable statistics).

The slowdown in the rate of increase of the catches since the 1960s is clearly due primarily to the limitations of the resources under the past practice of unregulated fishing. The question arises, can better fishery management increase the catches? The status of management has not been summarized, but a majority of the resources must have professional management in order to obtain optimum yields. About 60% of the major stocks appear to be fully or overexploited and the status of another 13% is unknown. Very few large stocks now appear to be managed in compliance with the somewhat idealistic provisions of the new LOS.

The prevailing management and development difficulties are extremely complex, with at least four major problems. First is the inadequacy of the conventional science for many important stocks. Few stocks satisfy completely the fundamental assumption of equilibrium needed for the mathematical models. They diverge into types[10] which have been classified in order of decreasing predictability as (a) the steady or predictable fisheries, (b) the cyclical fisheries, (c) the irregular stocks, and (d) the spasmodic stocks. Management of the last two is especially risky for the government policymakers as well as for investors in the industry. Improvement of the the ability to predict the abundance of most stocks will require long-term and expensive research.

Second is the lack of adequate catch and biological statistics on the stocks over enough time to provide a basis for prediction. This is especially true of stocks fished by large numbers of small-scale fishers who keep no records, and of all stocks on which fishing is not controlled by a single political entity.

Third is the transfer of authority and complex obligations of the new LOS to unprepared countries. The development of competent personnel and government organizations will take many years.

Fourth is the coordination between national governments and local communities. The national government may easily assert policies which may not correspond to local views and customs. In such cases, enforcement of laws may be impossible (See Part III).

The earlier projections of much larger production from the oceans were based on some very large resources that yet are little fished if at all. These include the Antarctic krill, a 5-cm-long crustacean that exists around much of Antarctica in huge quantities. Sustainable yields of 100 million tons have been estimated. Exploratory fishing began during the 1960s and expanded rapidly in the 1970s as the growth rate of conventional fisheries slowed. Production in the early 1980s rose to around 1 million tons, caught largely by vessels of the U.S.S.R.[11]

The problem has been largely a lack of product and market development. No widely acceptable human food has been produced. Indeed, it has been discovered that flouride in some samples far exceeded the quantity permissible under U.S. standards.[12] Most of the current catch has apparently been marketed as animal feed, but the costs of operations in waters so distant have discouraged nonsubsidized operations of most countries.

The krill has been a principal food of the baleen whales of the Antarctic, and efforts to protect the whales have resulted in political pressure to discourage krill production.

Other very large stocks include the mesopelagic finfishes and some oceanic squids that have been little fished. The mesopelagics are mostly small (10-15 cm) lanternfishes which exist at depths of hundreds of meters in several parts of the oceans in quantities of tens of million tons. Fishing techniques, products, and large markets for such finfishes and squids remain to be improved.

Other important possibilities for an increase in supply may be achieved through production of higher valued food products. This requires reduction of post harvest losses, development of human food products from catches and fish plant waste now directed to animal feeds, and salvage of fish now discarded, such as those from shrimp trawlers--which may be ten times the weight of the shrimp caught. Further, it will require special market development for low-priced products where they may be desirable.

The freshwater capture fisheries that have recently supplied about 10% of the world's commercial catches are unlikely to expand production beyond present levels. Many of the existing waters have been overfished for centuries, even millenia,[13] and they have been allocated to recreational fishing in many developed countries. Such market-exempt fisheries may be enhanced by the relatively expensive methods of stocking and habitat improvement when such measures are paid for through special license fees and are accepted by all people concerned. But these measures are unlikely to be feasible for small-scale market fisheries in most localities.

Correction of overfishing by small-scale market fisheries requires effective government policies and acceptance by the people concerned. Both are difficult to establish because of the technical complexity of the resources and the social changes that must be made.[14] Therefore, the catch trends from the freshwater (and lagoon or estuarine) capture fisheries probably will arise from the balance of two opposing tendencies. Decreases may be caused by the ever-escalating usage of water for other purposes and the deterioration of the habitat. Increases may occur in freshwater impoundments if they are planned and managed for increased production of fish. The prevention of damage or correction of past damage to the environment and planning for fishery values in new projects will be locally important, but it seems unlikely that established small-scale fisheries can increase their sustainable catches.

4.2.2 Fish Farming

Fish farming is an ancient activity that started when the supply of food or ornamental fish was not adequate. It has grown slowly over the centuries in fresh waters and sheltered coastal waters as practical people discovered methods of growing finfish, mollusks, and crustaceans. It began to receive scientific attention in the latter part of the nineteenth century and expanded in Europe and North America, where a major purpose was to stock finfish in natural waters for recreational fishers. Now, since the rapid expansion of fishing has ceased, it offers more promise.

The worldwide production statistics from fish farming are relatively poor, but the best available are summarized in Table 4.1. These show an annual rate of growth of 7.3% in the late 1970s. This happened in regions where it had already been widely practiced. In Africa, Latin America, and the Caribbean, where fish farming has not been traditional, there was little change.

The potential for future increases is indicated in part by the increases in production per unit area that are possible (see Section 3.4), and by the diversity of

Table 4.1
Aquacultural Production, 1975, 1980

	Africa	Latin America, Caribbean	Asia, Oceania	Europe	U.S., Canada	World total
1975	23	65	3584	817	131	4620
1980	12	67	5177	1138	188	6582

Source: FAO, 1984.[15] Amounts are in thousands of metric tons. Aquatic plant farm production of perhaps 3000 tons is not included.

opportunities. Fish farming practices range in sophistication from those that artisanal fishers can readily adopt to those requiring skills in water and stock management, product development, and distant marketing. Especially attractive for the former are simple ways of attracting wild fish or reducing competition among species and protecting them from predators in small multipurpose reservoirs.[16]

Intensive fish farming is a relatively sophisticated kind of farming that requires genetically amenable or domesticated organisms suitable for a farm environment that can be provided. It requires attention to the environmental, nutritional, and disease problems of each species being grown. These problems have been researched quite extensively during the past century for the species grown in the developed countries. But research on warm-water fish farming suitable for developing nations has been diffuse, muddled, and only moderately successful, even though a few of these countries produce the great majority of the world's farmed fish.[17]

The steps that have been so successful for increasing agricultural production should help fish farming. Research is essential, especially directed toward species already domesticated or known to be amenable, on "seed, health, feed, and genetic improvement in that order, and applied research on total culture systems."[18] Such research can be assisted by international cooperation and exchange of information, but it must be directed especially at the solution of local and regional problems and it must be accompanied by extension services for the farmers.

After farming systems have been proved, expansion of fish farming can be incorporated into the planning for water and agricultural development. Capture fisheries can be transformed into extensive type fish farms in fresh waters and estuaries.[14] Even in the Indo-Pacific region, where a major part of the world's fish farming already operates, the suitable area for fish farms has been estimated to be ten times as large as the existing farms.[19] Parts of irrigation projects, where land is too wet or heavy with clay for field crops, may be excellent for intensive type culture of fish. And fish may even be desert animals, where salt water from underground or oceanic sources is available for farming them.[20]

Many systems are already in use for farming high-priced products, such as tropical shrimp, freshwater shrimp, salmonids, and other species that can be fed artificial diets. These will undoubtably be expanded where water is suitable and low-cost feed is available. For example, Norway, one of the world's top ten fishing nations, in 5 years has tripled the production of salmonids farmed in its fjords. It is reported to be anticipating "explosive" further development that by the year 2000, may employ twice as many people as its entire fishing industry does now.[21] In addition, sea ranching of the anadromous salmonids is beginning in several countries.

And more possibilities are over the horizon. Artificial upwelling of nutrient-rich ocean water might be used, although its practicality has not been proved. Heated water from industrial cooling operations or geothermal sources may enhance the production of fish farms in the winter.

Now that the great expansion of ocean fishing is over and fish prices are rising, the potential for fish farming may be exceptional. The rate of increase from 1975 to 1980 of 7.3% annually is greater than the rate of increase of the ocean fishing during the expansion era of the 1950s and 1960s. If this continues and the additional potential begins to be realized, fish farming will soon become a much larger source of food.

4.3 COMPETITION WITH OTHER WATER USE

The freshwater fisheries compete for water required by domestic, industrial, and agricultural use. Such use changes the quantity and quality of the water and the structure of the aquatic environment as the water is stored, transported, consumed, heated, and polluted. The water requirements are about 30 m^3/person/year for domestic use and 400 m^3/person/year for agriculture. The water required by industry varies immensely from 20 to 40 m^3/person/year in the developing countries to more than 2000 m^3/person/year in the U.S.[22]

The saltwater fisheries also compete with the water required for urban and industrial use in harbors and are affected by the changes in water quality and structure of the environment associated with the upstream use of fresh water. This may have an impact on the many saltwater species that need sheltered estuarine waters at some stage in their lives.

The predominant mode of water management is damming in order to store water for later use, to avoid the damage and wastage of water due to floods, and to control water flow for hydroelectric power and navigation. Most of the rivers near population centers have been developed considerably, especially in China and Southeast Asia. The water development in China has involved construction of some 70,000 dams and reservoirs (mostly very small) over the past 30 years.[22] The least developed are the Amazon and Congo, which flow through inhospitable rain forests, and the large rivers, which flow into the Arctic Ocean.

Modification of government water policies is complex in all governments because of the multiple uses, the great value of many large-scale developments, and the numerous government agencies involved. The needs of the wild fish resources may be too complex technically to accommodate, and the impact of alternative projects too difficult to predict. Avoidance of damaging pollution is extremely important but frequently overlooked in hastily planned development. Also important is the need to include in the planning process the simple measures that avoid damage or actually improve the environment for fish. One example is provision of both low- and high-level discharges from reservoirs to control the water quality, both in the reservoir and downstream. Others are leaving brush or trees uncut in areas to be flooded in order to provide shelter for fish and constructing simple devices to enhance spawning for certain species.

Pollution in many drainages continues to increase, however, and much of the water development planning has failed to include consideration of either the potential damage or potential benefits to the fisheries.[23] A change toward multipurpose planning is desirable but it will not be easy.

4.4 IRRATIONAL FISHING

The rational use of a fish stock can be defined as obtaining optimum socioeconomic benefits while maintaining the size of the stock near its maximum level. Unfortunately, when market fishers have free access to the fishing, they eventually reduce the stock well below its level of maximum yield or, if restrained by government, they overinvest in an effort to increase individual shares and fail to optimize their benefits. This is irrational fishing which is commonly judged by its economic impacts.

Fishers from the earliest times have recognized the need for some restraint in their fishing. The earliest restraints were protection of eggs, young fish, and spawners. These commonsense restraints persist to this day as minimum size limits for individual animals or net meshes, as closed spawning seasons and spawning areas, or occasionally as maximum size limits and complete protection for mature females. The concept of protecting brood stock and young is, of course, basic to husbandry, and, in many cases, the very small fish are not worth seeking.

Each unit of a fish resource is limited by its environment and each has a limited harvestable portion. That portion is renewed by reproduction of adult individuals and growth of the younger. If it is not fished, it will reach a maximum population, varying according to environmental changes. If it is fished, it will be reduced in abundance, but doing so will reduce competition and allow more young to survive and the survivors to grow more rapidly--up to a point. At the point where catches equal the improved opportunities for survival and growth, the sustainable catches are at a maximum. Greater catches in any reproductive cycle will result in a reduction of future abundance, and then greater fishing effort will not result in greater catches.

Adherence to this simple and well-known ecological principle is easy in management of the nonmarket fisheries where fishers suffer no personal hardship with smaller catches. Government can make the fishing less efficient in proportion to the numbers of people who exercise their right to go fishing. Each person can have fun with perhaps only one fish instead of the thousand fish that each commercial fisherman might catch. They can, in some fisheries, have fun even when they release all of their catches alive. Furthermore, the economic impact of the supporting services for recreational fishers tends to increase as the catches are divided among more fishers. Thus, the economic importance of the recreational fisheries may be little affected by inefficiency.

But economic disaster occurs in market fisheries, even though they are managed, as more fishers exercise their public right. Troadec (24) describes the reasons as follows:

-there is often a delay of several years between the feasibility and investment analyses and the average operational life of ships...;
-each potential investor makes his decision independently, and therefore without being able to take account of the decisions made at the same time by his rivals;
-investments not justified by the size of the resource are often decided following periods where the natural abundance of the stock is above normal;

-the mobility enabling fishermen to enter the fishery is often greater than that for leaving it;...
-once the boat has been purchased it is easy to overlook its depreciation costs in estimating the real production costs;
-faced with the immediate social consequences of a reduction of means of fishing, which nevertheless is desirable in the long term, the fishery administrations are often inclined to subsidize the replacement of old boats...thereby perpetuating inefficient operations and inappropriate use of national financial and human resources.

The disaster is worsened by the usual fishery management policy focused on conservation of the resources, which commonly is accomplished by reducing the efficiency still further through gear restrictions, catch quotas, and other measures.

This generates what I call the abysmal cycle. The fishers still compete for a larger share of the catch by better gear, larger and faster boats, and better knowledge of the fish resources. They and any newcomers to the fishery still invest more money after good years, only to suffer even more in the poor years. The poor years generate need for government help. The next good year finds even more fishers competing, more government reduction of efficiency, and even less economic ability to withstand the next poor year. And so on. The industry, fishers and processors alike, teeters on the abyss of bankruptcy.

This problem of economic irrationality in the market fisheries is pervasive in small-scale and large-scale fisheries, in developed and developing countries--with the exception of some small island countries that have not yet expanded their fishing to the optimum sustainable yields.[25] In the small-scale fisheries of developing countries where fishing is an important rural occupation, the fishing villages have frequently grown long ago to a size limited by the food and employment furnished by the fisheries. In Canada, the United States, and most of the coastal countries of western Europe, many of the largest fisheries have overinvestments in vessels and processing facilities, and an increasing proportion are being subsidized by their governments. Such fisheries are economic drains, not economic benefits (see Chapter 9).

4.5 NEW GOALS

The future course of the world's fisheries is now clearly in the hands of national governments. They now have authority over virtually all of the capture fisheries of the world, freshwater, estuarine, and marine. But effective exercise of the authority requires a blend of technical and political skills that few governments have achieved.

In the market fisheries of all waters, governments must have twin goals: fishery management for conservation and fishery development for socioeconomic benefits, both of which require rational fishing. If the present, almost total preoccupation with conservation persists, the market fisheries will not contribute to economic development. Indeed, they will probably continue to require extensive government subsidies.

Management and development of the market fishing will be even more difficult for the stocks which must be shared with recreational fishers. The latter will insist on conservation and will probably have more political strength than the market fishers.

In the freshwater and estuarine fisheries, governments must protect the environment and divide the use of water among the many users.

In the small-scale fisheries of all countries, especially those that are developing, governments must balance the need for food production against the need for employment in fishing. Many of the stocks used by these fisheries are overfished, and introduction of a vessel or gear that increases efficiency tenfold for one fisherman can destroy the employment of nine.

Perhaps the most difficult goal for governments will be finding the proper balance between national and local concerns. The national government must control economic policy and relationships with other countries sharing the same water or fish stocks. It must also find ways to deal with the unique problems of each of its local fisheries.

Perhaps the easiest goal for governments will be the encouragement of fish farming. The farmers can own their fish and control their production in their own best interest.

NOTES AND REFERENCES

1. M. B. Schaefer and D. L. Alverson, World fish potentials. *In* DeW. Gilbert, ed., "The Future of the Fishing Industry of the United States." *Univ. Wash. Publ. in Fish., New Ser.* *4*:81-85 (1968).
2. R. I. Jackson, Famine and the world fisheries. *Trans. Amer. Fish. Soc.* *100*(1):151-158 (1971).
3. United Nations Environmental Programme (UNEP), "Environmental Impact Assessment; A Tool for Sound Development." *Ind. Envir., Guidel. Ser.*, (Spec. Iss. 1), UNEP, Paris, 20 pp., 1980.
4. W. T. Burke, Extended fisheries jurisdiction and the new Law of the Sea. *In* B. J. Rothschild, ed. "Global Fisheries: Perspectives for the 1980s". Springer-Verlag, New York, pp. 7-49, 1983.
5. U.S. Dept. of State, Third United Nations Conference on the Law of the Sea: Informal single negotiating text and text on settlement of disputes. Office of Law of the Sea Negotiations, Washington, D. C., May 9, 1975.
6. United Nations Environmental Programme, "Guidelines for Assessing Industrial Environmental Impact and Environmental Criteria for the Siting of Industry." *Ind. Envir. Guidel. Ser.*, *1*:1-105, UNEP, Paris, 1980.
7. U.S. Government, "The Global 2000 Report to the President." Fisheries Projections from Vol. 2, reprinted in *Fisheries (Bethesda)*, *6*(5):30-37 (1981).
8. J. P. Wise, The future of food from the sea. In J. L. Simon and H. Kahn, eds., "The Resourceful Earth: A Response to Global 2000." Basil Blackwell, pp. 113-127, 1984.
9. M. A. Robinson, Trends and prospects in world fisheries. *FAO Fish. Circ.*, (772):1-25 (1984).
10. J. F. Caddy and J. A. Gulland, Historical patterns of fish stocks. *Mar. Pol.*, *7*(4):267-278 (1983).
11. J. K. McElroy, Antarctic fisheries: History and prospects. *Mar. Pol.*, *8*(3)239-258 (1984).
12. T. Soevik and O. R. Braekkan, Fluoride in Antarctic krill *(Euphausia superba)* and Atlantic krill. *J. Fish. Res. Board Can.*, *36*:1414-1416 (1979).
13. W. Radcliffe, "Fishing from the Earliest Times." John Murray, London, 478 pp., 1921.

14. J. M. Kapetsky, Some considerations for the management of coastal lagoons and estuarine fisheries. *FAO Fish. Tech. Pap.*, (218):1-47 (1981).
15. FAO, A study of methodologies for forecasting aquaculture development. *FAO Fish. Tech. Pap.*, (248):1-47 (1984).
16. Z. Song, Manual of small-scale reservoir fish culture. *FAO Fish. Circ.*, (727):1-18 (1980).
17. R. S. Pullin and R. A. Neal, Tropical aquaculture: Need for a strong research base. *Mar. Pol.*, *8*(3):217-228 (1984).
18. FAO, Report of the ad hoc consultation on aquaculture research, Rome, 5-9 May 1980. *FAO Fish. Rep.*, (238):1-26 (1980).
19. H. R. Rabanal, Status of aquaculture fisheries in the Indo-Pacific Region. *In* T. Petr, ed., "Summary Report and Selected Papers Presented at the IPFC Workshop on Inland Fisheries for Planners, Manila, The Philippines, 2-6 August, 1982." *FAO Fish. Rep.*, Rome, (288):58-67 (1983).
20. H. Gordin, Aquaculture: Potential development. *In* P. G. Brewer, ed. "Oceanography: the Present and Future." Springer-Verlag, Berlin, pp. 347-361, 1983.
21. Anon., 1983 is Norway's best year for fish farming. *The Fisherman's News*, (Seattle, Washington), 40(19):2 (1984).
22. R. P. Ambroggi, Water. *Sci. Am.*, *243*(3):100-118 (1980). This is an issue devoted to economic development.
23. T. Petr, Resources development planning and management and man-made impacts on inland fish and fisheries. *In* T. Petr., ed., "Summary Report and Selected Papers Presented at the IPFC Workshop on Inland Fisheries for Planners, Manila, The Philippines, 2-6 August, 1982." *FAO Fish. Rep.*, (288):69-75 (1983).
24. J.-L. Troadec, Introduction to fisheries management: Advantages, difficulties and mechanisms. *FAO Fish. Tech. Pap.* (224):1-57 (1983).
25. B. C. Cleveland, National adjustments to changes in fisheries law and economic conditions. *FAO Fish. Circ.* (783):1-22 (1985).

PART II

FISHERY DEVELOPMENT EXPERIENCE

The central trend in the world's fishing for wild stocks of the commons has been from unregulated to governmentally controlled fishing; primarily for the purpose of conservation. This trend has been accompanied by unanticipated political and economic difficulties, which are associated with peaking of catches from the wild stocks, with rapid improvement in fish processing and marketing, with increased allocation of fish stocks to anglers, and with accelerating production from fish farming.

Major difficulties occur as market fishers, with investment in new equipment, may increase their catch per person by 10 to 30 times, and reduce the opportunities for poorer fishers to share in either the resource or the market. Further difficulties occur as anglers, who may far outnumber the market fishers, obtain preference in the allocation of choice species. Economic difficulties increase for the poorer market fishers who cannot continue their way of life without government subsidy.

The following chapters present an overview of the historical trends in fishery development, starting with its evolution over centuries and continuing with examples of trends during recent decades. Emphasis will be given to the different development problems of market fishing, fish farming, and recreational fishing.

Chapter 5

The Evolution of Fishing and Its Governance

The present condition of the world's fisheries is a consequence of the public use of fish and the public perception of how best to manage that use. Both the use and the perception have been evolving for millenia, and our understanding of the present predicament can be enhanced by a look at how we got here.

The ancient world depended heavily on fishing, as villages were frequently located near the waters. The fish were food at any time of year; fishing required little effort or gear; the catch was for personal use or trade in the local market. The fishing determined the existence of many villages.

The modern world depends little on fishing; as an economic activity it is seldom more than 1% of Gross Domestic Product (GDP) even in the largest fishing nations. Fishing is usually a very small part of national concerns, although it remains critical for some villages and a few small island nations.

The central feature of the public perception of the capture fisheries is that the resource is part of the commons, which everyone has a right to use but, as such, is subject to government control. This control has been a political process that has been influenced by scientific inputs only for about the past 40 years. It remains fundamentally a political process, in spite of great efforts on the part of scientists and passage of new laws requiring consideration of the science.

The public, as well as most scientific perceptions of fish farming, are similar to those of fishing. They are indeed similar in many biological respects, but the socioeconomic differences are such that they deserve different political attention.

This chapter is a brief summary of trends in government control of, and service to, the fisheries. It is intended to provide a background for our recent experience with fishery development that will be exemplified in the following chapters.

5.1 ANCIENT PRACTICES AND LAWS

Most of the shallow-water fish are among the easiest of wild animals to subdue, and they have been a major source of food for people who have usually lived near the shore in order to use them. An account of the fisheries of Sumeria (near the Persian Gulf), based on archives dated about 2300 B.C., has revealed something of the importance of fish as food and as religious objects during this time.(1) Several hundred fishers were employed by the temples (i.e., the government); they were second in number to the peasants, and outnumbered the herdsmen. They fished from shore or operated boats and caught about 90 species of fish which were preserved in several different ways. The fishers were organized in guildlike organizations each with a leader who was responsible for delivery of the catch, its quality, and its transport. The fishers formed separate groups which fished different waters, often with different kinds of gear.

The fishers were taxed and overseers collected the taxes. Ponds were built and stocked as fish culture was promoted. Poaching was a crime to be severely punished.

The fishing of Sumeria was organized around the religious beliefs (i.e., the laws). The fish were regarded as food, but other nearby peoples had very different beliefs and regarded some fish sacred to a deity as taboo.

The Romans adopted more formal laws, of which one established a principle that persists to this day in many countries. Fish and wild animals were placed in the category *Res Nullius:* natural things belonging to no one but which become the property of the person who first captures them.(2) Roman law also provided that the sea and the public rivers were not capable of individual ownership and no citizen could be prevented from fishing in them. Private ownership, although ill defined, was permitted of parts of rivers, coves, and fish ponds, subject to approval of neighbors.

5.2 DEPLETION OF THE STOCKS

Depletion has always been with us. Even the Sumerians found that the harvests were not uniformly abundant, and fishers had to go into debt. Decrees were issued to protect the catch of fishers from appropriation by the overseers. Fishers were at times unable to supply the government, so fish ponds were built and their operation regulated.

Depletion has always had secondary effects spreading through the businesses and lives of peoples associated with the fisheries and into their country politics. Very likely it periodically reduced the food supplies of the Sumerians or their neighbors. Certainly it had begun to affect the fisheries of eastern North America at the time of the Revolutionary War. One of the first reports on U.S. fisheries, by Thomas Jefferson, then Secretary of State, details some of the problems of New England fisheries in the late eighteenth century.(3)

The cod fisheries off northeastern America became a magnet for European fishermen soon after the settlers arrived in North America, when they were also vital to the local peoples. Hundreds of fishing vessels sailed out of ports in France, Spain, Portugal, and eventually England in the sixteenth and seventeenth centuries. The coastal Americans depended heavily on the fisheries for employment as well as food, and along with the English, prohibited importation of foreign fish. By the middle of the eighteenth century, after a war with France, the English and Americans dominated the fishery.

France, in an effort to maintain its sea power, gave its fishermen subsidies (in the form of bounties), and also added duties to imported fish. After the Revolutionary War, the English gave subsidies to their fishing vessels. By then the U.S. fishers had lost their markets in the Mediterranean and Canada, found their produce dutied in France, and had competitors supported by subsidies. Jefferson posed the issue to Congress of whether to match duty with duty, subsidy with subsidy, and how the country might create more markets for fish.

Similar problems occurred in the whale fisheries. Several hundred European vessels were whaling by the end of the seventeenth century, and the Americans began whaling in the early eighteenth. The North Atlantic whaling between Europe and North America had developed to the stage of depleted stocks during the early eighteenth century, and the whaling had spread to the South Atlantic in the later years.

By the end of the Revolutionary War, England, France, Holland, and the United States were all trying to control supplies and markets for whale oil through duties and subsidies. Jefferson stated:

> Against prohibitory duties in one country, and bounties to the adventurers in both of those which are contending with each other for the same object, ours have no auxiliaries but poverty and rigorous economy. The business, unaided, is a wretched one....No hope can arise of their condition being bettered by an augmentation in the price of oil. This is kept down by the competition of the vegetable oils, which answer the same purposes... This brings us to the question, what relief does the condition of this fishery require?

He went on to argue that the U.S. whalers, with their logistic advantages, did not need bounties but should be freed from taxes and that improvements in foreign markets should be negotiated.

The depletion of such large ocean fisheries as the Atlantic cod and whales must have occurred long after the depletion of the much smaller freshwater resources throughout much of the world. These individually mattered little to the countries, but collectively they must have been widely depleted before men endured the hazards of going to sea. Records of government action are fragmentary until the nineteenth century when they abound with concern about the inland fisheries and describe efforts to alleviate the depletion.

In many parts of the United States, fishing restrictions were imposed soon after settlers arrived. About the middle of the nineteenth century, state fishery agencies began to be established, and, in 1871, Congress established the U.S. Fish Commission to look into the causes of depletion in salt waters. Fishing licenses were soon required as a means of financing the state agencies.[4]

Also discovered or rediscovered about the middle of the nineteenth century were the techniques of controlling the reproduction of trout and rearing the young.[5] This was soon extended to other species, and the practice of fish hatching and rearing spread throughout North America and Europe for the purpose of restocking depleted waters. Species were exchanged between North America and Europe, and Atlantic coast species were transferred to the Pacific Coast of the United States.

The restocking programs assisted in stemming the depletion of the fresh waters and rehabilitating them. But the major step in controlling the depletion was an expanding program of legislatively prohibiting market fishing in the fresh waters of Europe and North America. The fisheries of the fresh waters were mostly allocated to recreational use early in the twentieth century. Sportsmen were prohibited from using excess ingenuity such as nets and required to be licensed. License fees were commonly earmarked for the sport fishery programs.

The sport of fishing was philosophically linked to the sport of hunting, and the legal management of both sports was a continuation of the Roman precedent of *Res Nullius*. The aquatic and the terrestrial animals were both considered to be resources of the commons, which all citizens had a right to seek, and which could become the property of any person capturing them. But it was obvious that one person could not freely harvest either without restraint and leave animals for others. So the sports people were restricted increasingly, in order that their take could be spread among those who wanted to take part, and would leave sustainable stocks behind. The fishing restrictions have brought the sports catches per person down to less than a hundredth, or even less than a thousandth, of the commercial catches.

The legal precedent was passed down through the centuries and formed the principal basis for European and North American wildlife law.(6) It became directed at goals beyond restricting ingenuity to include privileges for the gentry, humane killing, prohibition of the sale of animals caught under sports licenses, and improvement of the habitat.

As the freshwater fisheries and the coastal fisheries became depleted or allocated to recreational use, the market fisheries turned increasingly to the sea. When one stock became depleted, the fishers went farther, fished harder, and improved their gear. This human striving had existed throughout the centuries but did little to deplete the fish resources of the ocean outside the territorial sea until the development of steam propulsion and the otter trawl in the latter part of the nineteenth century.

This accelerated the depletion of the resources off northwest Europe, many of which were already subject to mildly restrictive regulation based on the commonsense husbandry of protecting spawners and preventing wastage of young fish.(7) Salmon were to be allowed to spawn in the rivers and their young return to the sea. Minimum mesh sizes had been prescribed for nets and minimum sizes had been required for fish to be sold.

Such regulations were, however, largely unenforceable and many were repealed as the rights and needs of fishers were given priority. T. H. Huxley eloquently stated the doctrine at the International Fisheries Exhibition in 1883 [as quoted in (7)]:

> Every legislative restriction means the creation of a new offence. In the case of fishery, it means that a simple man of the people, earning a scanty livelihood by hard toil, shall be liable to fine or imprisonment for doing that which he and his fathers before him have, up to that time, been free to do.
>
> If the general interest clearly requires that this burden should be put upon the fishermen--well and good. But if it does not--if, indeed, there is any doubt about

the matter--I think that the man who has made the unnecessary law deserves a heavier punishment than the man who breaks it.

Clearly, it was necessary to have better knowledge of the effect of fishing regulations if laws were to be accepted. Leading scientists from several European countries sought an international mechanism for collaboration on study of the complex problems, and several governments officially formed the International Council for the Exploration of the Sea (ICES) in 1902.(8) The fishing countries collaborated extensively on the scientific research, but Huxley's doctrine prevailed for another half a century.

The depletion went on and on; the bad times recurred more often; the fishers went farther to sea when they could no longer make a living on the nearby grounds. The stocks that were not depleted were saved temporarily by a limited market. The tendency toward depletion seemed inevitable and led a leading English fishery scientist, Michael Graham, to assert The Great Law Of Fishing--that fisheries in which the effort is unlimited become unprofitable.(9) This occurs despite all of the conservation efforts (see next section) to protect spawning fish or young fish or to limit catches. The ingenuity of the fishers using a resource of the commons has no limit.

Since then, as the examples in the following chapters will show, the Great Law of Fishing has been operative except on the few stocks not yet fully explored or accessible.

5.3 FISHERY MANAGEMENT

Fishery management has been a government function since before the time of the Sumerians, but until the mid-twentieth century, it was a political process shaped largely by the short-term interests of government rulers and the fish businesses. As depletion of the resources continued, however, other people became concerned, including those who wanted to enjoy fishing without profit and those who saw healthy fish as a reflection of good water quality. Maintenance of the fisheries in many countries became a part of the conservation ideal. Conservation of fish implied wise use, avoidance of waste, maintenance of the quality of the aquatic environment, restoration of depleted stocks, and perpetuation of the resources.

Achievement of these goals required sound information on the condition of the stocks and the effects of fishing on them. Scientific research programs were started during the last half of the nineteenth century and gradually gained political acceptability during the first half of the twentieth century. A Russian scientist(10) advanced the fundamental concept in 1918 that a year class (cohort) of fish grew to a maximum weight and then declined, but his paper received little attention until the 1930s. Then, several European scientists moved the concept further along to the idea that catch from a stock of fish would increase with more fishing effort up to a maximum and then decline as effort continued to increase. Scientific determination of the maximum sustainable yield and regulation of the fishing accordingly came to be seen as a solution to the pervasive depletion problem.(11)

A major impetus for adoption of the science-based conservation principle to the market fishery resources followed its application under the Northern Pacific Halibut Treaty between the United States and Canada that was ratified in 1924. The stock was studied, the fishing effort reduced, and changes in the relationship between the fishing effort and the subsequent catches became predictable. This information was used to frame the regulations which became widely accepted by the halibut businesses in the 1950s.

After World War II, the concept of fishing regulation for the purpose of conservation and according to the results of scientific study was adopted by many national fishery agencies and made a part of numerous international treaties.(12)

But international regulation of market fishing was deeply intertwined with issues of sovereignty over territorial seas, use of other marine resources, and freedom of navigation. More importantly, the fish businesses distrusted or remained unfamiliar with the scientific arguments, and Huxley's doctrine prevailed. International application of scientific findings to the fishing off northwest Europe was delayed more than half a century after the formation of the ICES until agreement on the Northeast Atlantic Fisheries Convention of 1959. The scientific conservation principles were pressed, however, at meetings preparatory for a new LOS in Rome in 1955 and in Geneva in 1960. A consensus on a critical provision for national authority--an EEZ 200 miles wide--was not reached until a meeting of the third Law of the Sea Conference in 1974. The Convention was not adopted until 1982 (see Section 4.1.), but most coastal countries took unilateral action to claim EEZs before 1979.

The market-exempt (recreational) fishers, however, consented much more easily to science-based regulation for the purposes of conservation. Conservation became a creed, implying respect for and care of the environment. The presence of fish indicates the quality of the water, which in turn is usually associated with the pleasantness of the location.

The fresh waters of the world had long suffered from depletion of the fish stocks, and most of the developed countries restricted or forbade market fishing in fresh waters during the first half of the twentieth century or earlier. The fishing was allocated to recreational fishers who value the pleasure associated with fishing as much or more than the catch itself. These fishers now comprise large fractions of the population in the developed countries (about 25% in the United States). They strongly support government programs of research on the fisheries and on the aquatic environment. They expect their governments to preserve the ancient right of everyone to go fishing and expect that right to be fairly distributed among those who want to fish. They pay much, if not all, of the costs of government management through special license fees or special taxes on equipment. The exercise of rights and perceptions of fairness are to be preserved regardless of how few fish may be caught (see Chapter 8).

The social, economic, and political dimensions of the recreational fisheries in most developed countries are already as large or larger than those of the market fisheries.(13) Their management has been almost entirely a national responsibility, and nations are just beginning to discover common interests and organize to exchange information on experience and research.

The recreational fisheries, however, have an entirely different business dimension. If the amenities of the environment and the political perceptions are maintained, the economic value can be almost independent of the size of the catch. The management practice of dividing the catch among those who want to fish, which is irrational in the market fisheries (Section 4.4), usually has little effect on the recreational fisheries. In fact, spreading the catch among more people who spend their money on travel and equipment will generate more economic benefits--a result exactly opposite to spreading the catch among more market fishers.

Fishery management by governments extends also to certain aspects of fish farming. Some of the private farming depends on brood stock or seed from wild resources, and some farm sites in public waters (e.g., oyster farms or fish cages) must be used with permission from the government. In such cases the government may need to settle differences between the fishers and farmers using the same resources or resource areas. Other public concerns arise because aquatic farm animals may be transported alive from place to place and may carry diseases or pests. The animals themselves or their diseases and pests may escape to cause trouble for other fish farmers or ecological changes in native populations of fish. Some of the farm species may be the same as native wild species, the sale of which may be prohibited. Governments usually intervene to control such problems.

5.4 FISHERY DEVELOPMENT

The goal of fishery development is, in its broadest sense, to use the fisheries to improve the well-being of people in a country. The objectives may include increased production, increased supply of desirable protein food, increased recreational opportunities, increased export earnings, increased employment, and increased earnings for fishers. Since such objectives frequently conflict one with another, it should be noted that it is impossible to maximize more than one objective at one time.

The great challenge to governments is how to intervene in the fish businesses in order to achieve any of the objectives, while continuing to achieve conservation of the resources through fishery management. Although the governance of fishery development and management is linked through concern for the resources, the practice of fishery management has commonly inhibited development of the market fisheries.

Whereas government fishery management for conservation became widely accepted about the middle of the twentieth century, market fishery development was left almost entirely to the business sector.[14] Worldwide attention to market fishery development began with formation of the FAO in the late 1940s. The FAO began collection of worldwide statistics on the catches, commodities, and trade. These revealed the rapid expansion of the fisheries in the 1950s and 1960s that occurred because of the new technology, the discovery, and the exploitation of the major fishery resources in the world's oceans. Along with expanding fisheries came expanding scientific studies of the oceans and the fishery resources.

Another incentive for fishery development came after the international consensus in 1974 on a 200-mile EEZ, and all coastal countries anticipated further economic development from the fisheries which they could then control. National fishery assistance projects proliferated, many of them in the developing countries with contributions from developed countries or international agencies. In addition, because nations must agree with their neighbors on fishery management or development, special efforts were made to frame effective international fishery conventions.

Most of the fishery development efforts by governments have focused on the oceanic market fishing, but recreational fishing and fish farming are receiving increased attention. Recreational fishery development, that revolves around the concepts of conservation, is widespread in the western countries, and it is becoming important among less developed countries as part of the development of tourism. Fish farming is developing rapidly as a part of the agricultural business sector where the development problems are quite different than those of the market fisheries on wild stocks.

Most coastal countries have, therefore, formulated fishery development plans, but achievement of successful fishery development has proved to be an especially difficult government task.(15) The reasons will be elaborated in much of the rest of this book, but they include the following:

1. The technological model of market fishery development that prevailed until about 1970 is no longer appropriate because of scarcity of resources and acute social problems in fishing villages.
2. The scientific model for management of the resources requires sound statistics on catches, expensive scientific programs, and effective enforcement, all of which are difficult for most countries,
3. The freshwater and coastal aquatic environments are used for many conflicting purposes which must be balanced by governments.
4. Nations must collaborate in fishery management and development because many of the resources cross their boundaries.

NOTES AND REFERENCES

1. R. Kreuzer, The cradle of sea fisheries. *In* E. M. Borgese and N. Ginsburg, eds., "Ocean Yearbook 1." Univ. of Chicago Press, Chicago, pp. 102-113, 1978.
2. W. Radcliffe, "Fishing from the Earliest Times." John Murray, London, p. 231, 1921.
3. T. Jefferson, Communication on fisheries to the House of Representatives, February 2, 1791. Reprinted in *Fisheries (Bethesda) 1*(3):15-20, (1976).
4. H. Clepper, ed. "Origins of American Conservation." Ronald Press, New York, 1966. See Chapter 4, pp. 57-89, Fisheries and aquatic resources: Lakes, streams, and other inland waters by R. H. Stroud, Coastal and marine waters by C. P. Idyll.
5. W. H. Fry, Transl. and ed., "A Complete Treatise on Artificial Fish Breeding." Appleton, New York, 188 pp., 1854.

6. T. A. Lund, "American Wildlife Law." Univ. California Press, Berkeley, 179 pp., 1980. See Chapter 1, pp. 3-17, English wildlife law before the American Revolution.
7. M. Graham, Concepts of conservation. *In* "Papers Presented at the International Technical Conference on the Conservation of the Living Resources of the Sea, Rome, 18 April to 10 May 1955." United Nations, New York, pp.1-13, 1956.
8. A. E. J. Went, Seventy years agrowing: A history of the International Council for the Exploration of the Sea. *Rapp. P.-v. Réun. Cons. Int. Explor. Mer.* (165):1-252, (1972).
9. M. Graham, "The Fish Gate." Faber and Faber, London, 199 pp., 1943. See Chapter 13, pp. 140-155.
10. F. I. Baranov, On the question of the biological basis of fisheries. *Nauch. Issled. Ikthiol. Inst., Izv. 1*(1):81-128 (1918). (In Russian).
11. E. S. Russell, "The Overfishing Problem". Cambridge Univ. Press, London and New York, 130 pp., 1942.
12. D. M. Johnston, "The International Law of Fisheries: A Framework for Policy-Oriented Inquiries." Yale Univ. Press, New Haven, Connecticut, 554 pp., 1965. See Part Three, Patterns of Conservation Authority. Early fishery conventions in which science-based conservation played an important role include (in addition to the Halibut Convention) the Baltic Convention of 1929; the Convention for the Protection, Preservation, and Extension of the Sockeye Salmon Fisheries in the Fraser River System of 1930 (not ratified until 1937); the Northwest Atlantic Fisheries Convention of 1949; the Inter-American Tropical Tuna Convention of 1953; the International North Pacific Fisheries Convention of 1953; and the Northeast Atlantic Fisheries Convention of 1959.
13. J.-L. Gaudet, Introduction and Summary. *In* J. H. Grover, ed., "Allocation of Fishery Resources: Proceedings of the Technical Consultation on Allocation of Fishery Resources Vichy, France, 20-23 April 1980." FAO and Amer. Fish. Soc., pp. 1-7, 1982.
14. The business sector in most countries is the private sector but government operates the fish businesses in the countries with centrally planned economies and occasionally in others as parastatal businesses.

15. A. C. S. Diegues, Policies and strategies for fisheries development: Some relevant issues for developing countries. *In* Case Studies and Working Papers Presented at the Expert Consultation on Strategies for Fisheries Development, Rome, 10-14 May 1983." *FAO Fish. Rep.*, (295, Suppl.), pp. 1-15 (1983).

Chapter 6

Case Histories of National Market Fishery Development Experience

Essentially all of the market fisheries of the world have become subject to fishery development by governments because people are aware of their social and economic importance and have sought, or at least accepted, government intervention in their use. Development is a dynamic long-term process within which all of the private and public decisions about investment, employment, markets, technology, research, management, and policy pertain only to a moment. Further, these decisions are based on factors internal to the fisheries, such as perceptions of the fishers, state of the stocks and markets, and opportunities for investment and employment, as well as on factors external to the fisheries, such as other uses of the waters, need for food, alternative opportunities for investment and employment, and public perception of the role of fishing. Any one of these factors may impede or enhance the development process.

This chapter is an attempt to describe some historical trends in the development process since the 1940s. This period includes the era of technological expansion of the fisheries into the early 1970s and the first few years of widespread assertion of authority over a 200-mile EEZ. Although I include one village-level study, the experience of a few national governments in fostering development, either domestically or through international assistance, is the focus. Emphasis is given to policy and institutional problems, since these areas usually control the relative success of the programs and have had less scholarly attention than the scientific and technical problems.

The examples have been chosen because suitable published material is available and because they illustrate the problems of small-scale fisheries, the transition to large-scale fisheries, and the problems of the largest fisheries in the world. All of the countries discussed but one have, at one time or another, been included in the list of the ten largest fishing nations in the world.

The coverage of the examples is necessarily spotty because comparable material for different countries is difficult to obtain and because many of the policy actions are reported in the public press rather than in the scientific literature. And all of the examples have probably changed since the latest documentation. Such brief historical reviews may also blur the complexity of the problems, especially with regard to the relationship of national and local institutions.

6.1 A FUNDAMENTAL DIFFERENCE--INDIA'S KERALA COAST

India, with its long coastline, its large rivers, and its large population of poor people, has long sought to make maximum use of its fisheries. It ranked among the top 10 nations in fish production in the early 1950s, but most of its fishers were artisanal and most of its catch was taken in fresh waters or within 25 km of the coast.

The United States responded to a request for assistance with a limited development program which included introduction of 32 new vessels, provision of diesel engines and outboards, fishing nets and other equipment, a net-making plant, 12 ice-making plants, insulated or refrigerated trucks and railway cars, and two pilot fish meal plants. It also tried to improve fish pond operations and encouraged formation of cooperatives. The program operated from 1952 to 1962.(1)

Although Indian fish production expanded, the results of the program were disappointing. It was discovered that the large population of artisanal fishers had enormously diverse cultures and languages and limited ability to use the mechanized equipment.

During about the same period, the Norwegian government supported a learning project with extensive participation by social anthropologists. This focused on two neighboring villages on the Kerala coast--the southernmost 500 km of the west coast of India.[2]

The Kerala Coast at that time had about 100,000 active fishers who worked the most important resource of the area. The people lived in small villages scattered along the shores of the inlets and lagoons where they were protected from storms by the offshore sandbars. The fishing was conducted in the lagoons, inlets, and within about 8 km of the outside coast, with lines, cast nets, shrimp traps, or gill nets, operated from shore or canoes. The catches were sold (after giving away some to beggars, holy men, and poor children) to buyers (middlemen) who were highly competitive.

The two villages were chosen for special study because they typified the two major groups of fishers along the coast. One was Latin Catholic, descended from early Portuguese explorers; the other Hindu, in which the fishers belong to the untouchable caste.

The project, which continued for more than a decade in close collaboration with the Indian government, involved improvement of the fishing gear; motorization of the canoes, first with outboard then with inboard engines; and introduction of larger motor boats, up to about 11 m in length. Changes were made gradually, because each necessarily required change in the fishers lives. Bigger boats required a change in landing places, gill nets performed better when fished at night, larger boats required more financing (frequently by nonfishers), larger catches could be handled and sold better through a cold store, captains had to divide their time between the critical fish auctions and operation of a more complex boat, trucking services expanded, and opportunities came for some to invest in income-producing assets.

Many facets of the lives of the people in the two villages were investigated over the years as the project continued. The villages were only a few kilometers apart, and every effort was made to treat them as equally as possible.

Profound differences appeared to be related to the social backgrounds. The Catholics were prudent with money and were investors. They were entrepreneurs who saved much of their money and used it to advance their rank and status. They readily adopted the practice of night fishing. Many found private financing and bought boats without subsidies.

The Hindus, on the other hand, appeared to be resigned to their status. They were reluctant to fish at night. Larger earnings were spent on such things as jewelry or for entertainment. Few of them repaid the loans on the boats.

A cooperative sales organization and a cold store were started. By guaranteeing a fixed price, the cooperative got a monopoly on all fish sales and lost large sums of money within a few days, whereupon the middlemen resumed buying.

The buyers and distributors were at first regarded as gougers, but they were highly competitive among themselves and dominated the marketing throughout the life of the project. It eventually became clear that they provided an essential service.

The two communities drifted apart as political stresses emerged. Many Hindus illegally rented their boats to the Catholics. Hindu leaders sought excuses in uneven treatment by the project, and a leader complained that "our misfortune becomes more difficult to endure." At the termination of the project, it was praised by the Catholics, not by the Hindus.

The social scientists involved with the study concluded that it was probably impossible to anticipate such results, but that the social changes generated by the technical inputs deserved much more careful consideration in future projects.

Fortunately, the experiences of this project on the Kerala coast and others in India have led to more careful consideration of the social problems and more effective fishery development programs. The basic strategy requires recognition by governments that they must plan fishery development as a long-term process of social change (see Chapter 10).

6.2 AN EVOLUTIONARY PROCESS--SIERRA LEONE

The northwest coast of Africa, from the Strait of Gibraltar to the Gulf of Guinea, is a relatively rich fishing area. It was thoroughly explored by distant-water fleets from the major fishing countries of Europe, the U.S.S.R., Japan, and the Republic of Korea during the 1960s. Total production leveled out around 3.5 million tons during the 1970s, of which about one-third was taken by the U.S.S.R. Almost all of the fish stocks appeared to be fully exploited at that time.(3)

Some of the best fishing occurs off the westernmost tip of the continent, where the upwelling ocean waters are further enriched by large rivers and coastal lagoons. This area includes the countries of Mauritania, Senegal, The Gambia, Guinea-Bissau, Guinea, and Sierra Leone. These six countries are among the poorest in the world and, therefore, have expected important benefits from fishery development.

The local fisheries were almost entirely artisanal, using beach seines, gill nets, lines, spears, poisons, and simple traps. Most of the vessels were canoes without power. Yet fish were an important part of the food supplies, and as foreign trawlers began to operate, some of their catches were sold in the local ports.

After the consensus on a 200-mile EEZ in 1974, these countries sought additional food and economic benefits from the sea. Most of them soon extended their jurisdiction to 200 miles. They began to seek assistance in order to invest in vessels, cold stores, port facilities, and help for their artisanal fishers. They also began to negotiate with the countries sending distant-water fleets in order to license, control, and benefit from such fishing.

The historical fisheries of Sierra Leone were probably typical of much of the area. I have included a discussion of them here because a considerable amount of pertinent information is available and because I have had personal experience in the country. It is a country a little smaller in area than Scotland, a little larger than Sri Lanka. It has substantial rainfall, and agriculture is possible throughout its area. It is, therefore, well populated with about 3.5 million people, whose culture is typical of West Africa. It had a long experience under British administration and became a sovereign state within the Commonwealth in 1961. It has been a parlimentary democracy since, except for about one year of military government in 1967-1968. The present national government has respected the chiefdoms, of which there were about 150 in 1963. English is the official language, but the more than 20 ethnic groups use at least another dozen local languages.(4)

The early European visitors to the country found the local people knowledgeable about the fish and making extensive use of them.(5) Fishing was a family activity in which the men fished and the women processed and marketed the catches. Processing was mainly smoke-drying.

The fisheries were relatively unchanged until Fante fishers arrived from Ghana during the 1950s. These people have been among the more aggressive and daring fishers of West Africa. They introduced a more efficient ring net, larger canoes, changes in crew size, a new division of labor, and new canoe construction methods. These changes meant employment of nonfamily members, a need for capital investment, and increased costs. The Fante also improved smoke ovens.

By the 1960s, much new technology had been adopted, but social changes continued in the fishing communities.(5) Nonfishing boat owners (including women) became more prevalent. A group of master fishers (captains) emerged. Family relationships changed as one wife was unable to manage the processing and marketing, so the communities changed from predominantly monogamous to predominantly polygamous (Islamized).

The increased production led to creation of distant markets and sales to fishmongers who appeared with lorries. This was accompanied by another ethnic migration to the main fishing coast, Fulas from Guinea, who specialized in wood cutting and firewood supply.

Soon after independence in 1961, a political decision was made to repatriate the Fante fishers back to Ghana. After the decade of change stimulated by these industrious fishers, they left behind very different fishing communities, but communities that were still essentially artisanal.

The remaining local fishers, however, were unable to supply the expanded markets, so Sierra Leone began to import frozen fish from foreign trawlers operating off its coast. A small number of cold stores was established to handle the fish for inland markets. In addition, about a half dozen trawlers owned predominantly by expatriates had begun operation out of Freetown by the mid-1970s.(6)

During the late 1960s, the government began to seek assistance in fishery development.(6) The UNDP financed a project managed by FAO that was conducted from 1967 to 1972 to evaluate the pelagic resources. The International Development Research Centre (IDRC) of Canada provided money in 1973 to study the possibilities of cultivating mangrove oysters. An Italian consultancy firm in 1974 studied the feasibility of various ways of expanding production and suggested construction of port facilities, introduction of medium size trawl and seine vessels, and training of fishing crews. Another FAO mission in 1975 concluded that the fishery sector had considerable development potential. The mission suggested upgrading of the canoes; training of boatbuilders; training of captains, mechanics and extension workers; preparation for better fish processing and marketing facilities; and provision of fishery consultant services. An FAO/Norwegian Agency for Development (NORAD) mission in 1975 agreed on the favorable prospects for industrial fishing. The Commonwealth Fund also surveyed landing facilities in 1975 and recommended construction of a major fishery harbor. A U.S. Agency for International Development (AID) mission reviewed the potential for fish farm development in 1976.(6) And numerous other assistance projects were under discussion in 1976 with support from the Danish International Development Agency (DANIDA), the Canadian International Development Agency (CIDA), NORAD, the Ministry of Overseas Development (UK), and the FAO.

The government's Development Plan for the last half of the 1970s projected the annual growth rate of the GDP at 6.2%, of the population at 2.6%, and of the demand for fish at 5.1%. It anticipated that with the assistance projects, the country could supply its entire demand for fish by the end of the decade.

During this period, some of the fishing opportunities were given away. A treaty of friendship and fishing was signed with the U.S.S.R. in 1976, which allowed fishing by 16 large freezer-trawlers. Another treaty with the neighboring country of Liberia to the south allowed fishers from each country to fish in the other's waters. This appears to have had little benefit to Sierra Leone, because the wide continental shelf and enriching ocean currents were off Sierra Leone, not off Liberia. In addition to legal fishing under the treaties, vessels from other countries invaded Sierra Leone waters. The country had essentially no capability to enforce regulations in its 200-mile zone.

In addition, the government entered the fish business with a share in a large joint venture and with directed controls on fish transportation and fish buying from foreign vessels.

And what happened to the country's total catches? They increased from a roughly estimated 30,000 tons in 1970 to a peak of about 69,000 tons in 1975, almost immediately declined and remained at an annual level of about 50,000 tons through 1983.[7] This left the country about 30,000 tons short of its projected demand for fish.

Fish farming had been considered but not recommended.[6] The country has a high rainfall and abundant fresh waters, but it also has extremely porous soils that require special and expensive pond construction methods. In addition, extensive African experience with construction of farm fish ponds in several countries had uniformly failed because of social problems and inadequate project preparation. (In one country, my personal visit to a government experiment station designed to foster pond fish farming revealed that it was experimenting with crocodiles, not fish.) The project to culture mangrove oysters was technically successful, but reports in the early 1980s indicated that almost no people had seen a business opportunity in oyster culture.

The history of fishery development in Sierra Leone is one of failure due to the social complexity of the fisheries and the inability of the government to devise and conduct policies essential for the fisheries. The political power in the country remained from 1968 to 1985 in the hands of President Siaka Stevens, who maintained a reasonable political equilibrium. He nominated his successor, Major General Joseph Saidu Momoh, Commander of the country's armed forces since 1971, who received 99.85% of the valid votes in an uncontested election in October 1985.(8)

The country is heavily in debt and commercial loans are no longer available. In addition to large fish imports, it imports about one third of its rice supply, a situation which is attributed to the low prices received by the producers. About 75% of the country's economic activity has been estimated to be outside legal channels.(9)

Assistance to the fisheries has continued with changing emphasis. A project to assist with technical and social problems was receiving US$750,000 annually from the Federal Republic of Germany in 1983.(10) Nevertheless, the government officially argued in 1984 that "there is still a dire need for economic and technical help from donor countries and agencies. Without this assistance, our goals would be unachievable in the forseeable future."(11)

6.3 RESTRAINED BY THE EEZ--THAILAND

Thailand is a country close to good fishing grounds with a large internal market for fish products. Its land area is about 514,000 km^2; its population about 35 million; thus it is about the size of and has about the population density of Spain.

The Gulf of Thailand on its east coast is a shallow sea, easily trawlable throughout most of its area, and one of the richest fishing grounds in the world due to the large rivers that feed it. The Andaman Sea on its west coast is much less productive per unit area, difficult to fish because of its narrow continental shelf and rocky bottom, but still an important fishing ground. In addition, the riverine areas, large mangrove swamps, and coastal lagoons are highly productive of fish.

From the 1930s until about 1950, the annual fish production of Thailand was about 160,000 tons. Most of this was taken by artisanal fishers using bamboo traps, set nets, and purse seines operated from small boats. Most of the boats lacked power until after World War II, but by 1960 about half the 2,800 boats had power. The U.S. government had assisted by supplying engines and nylon gill nets.(12)

Then Thailand received assistance from the Federal Republic of Germany under a bilateral agreement of 1956 for technical and economic cooperation. Plans were prepared during the late 1950s for the introduction of bottom trawling, a fishing method unused in Thailand, but highly developed in Europe and North America. Trawling previously had been introduced in the Philippines where the principal German scientist had had experience with its development.(13) The first phase of the project operated during the summer of 1961, the second from 1962 to 1964.

Research cruises began in 1961, primarily with trawlers of about 20 m length constructed in Thailand. These were augmented by larger research vessels, one of which was supplied by Japan. By 1964, eight research trawlers were operating.

The trawlers were supplied initially with nets made in Germany, and the operations were supervised by master fishers from Germany. The crews of the research vessels, 280 fishers and several officers of the Thai Department of Fisheries, were trained in construction and repair of nets and the conduct of trawl fishing during the second phase of the project. In addition, the new trawling techniques were demonstrated aboard private commercial vessels in order to allow the owners to apply the experience promptly and directly.(14)

The research trawlers were used to systematically survey the bottom fish resources of the Gulf of Thailand. They soon produced estimates of the distribution and abundance of the many species of bottom fish. Eventually it was possible to estimate the sustainable yields. Such research was a major departure from the earlier hydrographic surveys and marine biological studies which had dominated the marine research in Thailand and several neighboring countries.

The results of the research and demonstrations led to extensive domestic investment in fishing vessels and shore facilities. The number of trawlers registered to operate in the Gulf of Thailand was 99 in 1960, 2026 in 1963 and 5834 in 1977.[12] The total number of motorized vessels throughout the country rose to about 19,000 in 1967 and to about 21,000 in 1976, whereas the number of nonmotorized vessels decreased from about 15,000 in 1967 to only 5000 in 1976. The fishing capability increased even more, since the average new vessel was much larger and more capable than the old; hence catches increased rapidly (Fig. 6.1).

The technical changes brought socioeconomic changes. As the new capital-intensive fishery grew rapidly, the fisheries became important in the national economy, amounting to about 3.3% of the GDP or about 10.3% of the gross agricultural product by 1977. They supported many jobs in the fishery supply, in vessel construction, in cold storage, and in fish marketing businesses. Fish exports rapidly increased reaching 133,000 tons valued at more than 3000 million Baht in 1976.

The technical changes also brought depletion of the nearby resources. The research trawlers of the Department of Fisheries caught about 290 kg/hr in the Gulf of Thailand in 1963 but less than 60 kg/hr in 1976. Many of the bottomfish resources in the Gulf were judged to be overfished as early as 1968, but no appropriate action to restrain fishing had been taken by the end of the 1970s. The potential for increased production from Thai waters was estimated in the mid-1970s to be only about 30% of the current level, most of which would have to come from pelagic species and all of which would have to come from species more difficult to catch or less desirable in the market.

The freshwater fisheries and fish farming in Thailand also increased but not as rapidly as the commercial fishing.[15] The total production approximately tripled from the level of the 1950s. Shrimp and prawn farming increased rapidly in the 1970s to a level of about 7000 tons in 1978.

Many of the artisanal fisheries continued to exist as a rural occupation linked to farming in communities with low income and high unemployment.[16] The number of

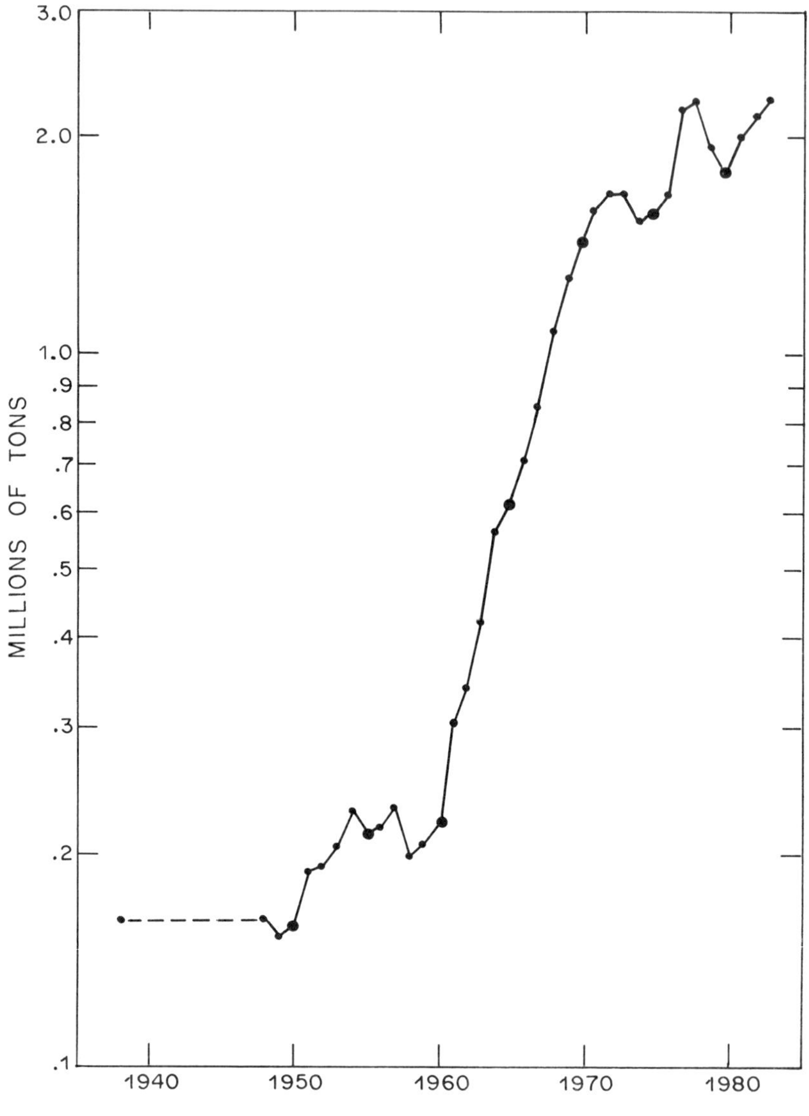

Fig. 6.1. Fish catches by Thailand. Source FAO "Yearbooks of Fishery Statistics: Catches and Landings, 1948-1983.[7]

artisanal fishing households decreased only slightly during the early 1970s, and by 1980 the government was trying to help them by promotion of fish farming.

Unfortunately, the market fishing sector could provide few opportunities for the large numbers of such people. Their fishing gear was subject to accidental damage by the trawlers and their traditional stocks were depleted. They had few options in the poor villages in which they lived and little if any mobility in their selection of fishing areas. They had less political and economic strength than the trawl fishers. Any solution to their social problems had to come from outside the fishing sector, whereas Thailand already had extensive unemployment and underemployment.(17)

The technical changes in the ocean fisheries also brought about international political problems. Investors supported larger and larger boats, many of which could fish far beyond Thai waters where they needed to go because the abundance of fish in Thai waters had decreased greatly by 1970. By the mid-1970s they were fishing up to the 3- or 12-mile territorial limits of many countries, including India, Bangladesh, Burma, Indonesia, Malaysia, Kampuchea, and Vietnam. And, because few of the coastal waters were patrolled, the venturesome Thai fishers also poached in other countries' waters.

As a distant-water fishing nation, Thailand resisted the adoption of 200-mile EEZs and did not recognize the legitimacy of other countries' declarations. But in February 1981, Thailand gave in and declared its own EEZ.(18) This was expected to cause major changes in the fish business, because Thailand would lose about 700,000 km^2 of fishing area and might suffer a reduction of 600,000 to 800,000 tons of marine catch.

Nevertheless, the Thais fished harder and negotiated more. The Gulf of Thailand became more overfished than ever, causing acrimonious conflicts between domestic artisanal and large-scale fishers.(19) Conflicts with foreign countries also increased because the ocean borders were not marked nor could they be agreed upon with several neighbors. Thousands of Thai fishers were arrested and jailed in Vietnam, Kampuchea, and Burma. Many vessels were seized and the owners were heavily fined or had their vessels confiscated. Joint ventures were arranged with several countries including Bangladesh and China, but the era of expanding ocean fishing from Thailand appears to be over.

The rapid development of the Thai fishery was possible because of the ready acceptance of the technical and social changes by the Thai government after the introduction of trawling.[20] The changes accounted for a tenfold increase in production in about a decade. Now the major ocean fishery problems of the government are how to create the domestic and international policies that will match the fishing effort to the productivity of the resources in a socially acceptable way that does not require continuing government subsidy. The major problems of the coastal artisanal fishers are socially even more difficult and will require that solutions be found in fish farming or other activities outside the capture fisheries.

6.4 A FICKLE CURRENT--PERU

Peru's ocean area is one of the most productive fishing grounds in the world. Yet it made little direct use of its fish until it built an extraordinary export market.

Its land area of about 1.3m km^2 is much larger than that of any western European country, but it is sparsely populated with only about 14 million people. Its agricultural possibilities are limited by its mountainous terrain and coastal deserts. Surprisingly, in the early 1940s the Peruvian people consumed few fish--only about 2 kg/person/yr. The total fish landings then were in the range of 25,000 to 35,000 tons annually.[21]

The ocean off Peru provides the classic example of an upwelling current that supports immense stocks of fish. The prevailing winds drive the surface water offshore and thereby bring the rich deep waters up into the photosynthetic zone where plankton grow and support fish. The coastal surface waters are cool, 14-16°C, whereas 150-200 km offshore the temperature of 20-25° C is more typical of the tropical latitudes.

The usual ocean condition off Peru, however, is occasionally overridden by major changes in both atmospheric and oceanic circulation of much of the Pacific Ocean. This has recurred irregularly but with an average frequency of about once each decade for many years. Its major effects have usually occurred about Christmastime and hence have been named "El Niño" with reference to the birth of Christ.

The most abundant fish in these waters is the anchoveta, *Engraulis ringens*, which is better adapted to the low temperature and to the low dissolved-oxygen levels of the coastal waters than any other species. Consequently, large populations of migratory predator fish such as tunas usually find the waters undesirable and do not consume the anchoveta in major numbers.

The anchoveta are commonly caught at sizes from 10 to 20 cm long when they are about 6 months to 3 years old. They are schooling fish, readily captured by fine mesh purse seines within a few hours run from shore by fishing vessels.

For about a century prior to 1950, the anchoveta was used indirectly, since it was the principal food of the sea birds which roosted and nested on the almost rain-free rocks and islands along the coast. The immense deposits of bird guano were heavily exploited and mostly depleted during the last half of the nineteenth century when the guano was exported to Europe and North America for fertilizer. This led to government intervention in 1909 with the creation of the Compañía Administradora de Guano (CAG). The CAG protected the sea birds and controlled the guano harvest thereafter.(22)

Other species of fish were caught by the artisanal fisheries for local food. The principal species taken in 1940 were bonito, mackerel, sharks, several species of the sea bass and croaker families, and several species of crustaceans and mollusks. Most of these were consumed fresh; most of the balance were salted and dried. Large quantities of edible seaweeds were also harvested.

The occupation of fishing was reserved to Peruvian citizens, almost all of whom were native Indians. Fishing vessels were mostly small sailing craft, log canoes, or reed boats. Only 87 out of about 2400 boats had motors in 1940.

The major commercial activities in the fisheries were apparently the harvest of sea lions, of which about 80,000 were taken annually for their leather; the beginning of a fish liver business in which the livers of bonito, tuna, and sharks were saved for their vitamin content; and a small export of frozen fish.(21)

The United States assisted with fishery development in Peru by sending a mission in 1941 which fully described the fishery, conducted exploratory fishing, and made recommendations for advancement of the industry.(21) Apparently this was related to a Declaration of Panama in 1939, in which the United States and the Latin American nations insisted that their waters should be free from war-related activities to a reasonable distance from their coasts. In effect, they declared a *de facto* sovereignty when they were threatened.(23)

In 1945, after the war, President Truman issued a Proclamation in which he claimed for the United States the right to establish conservation zones outside the territorial sea. In these zones fishing would be controlled independently by the United States, or if foreign fishers were involved, through agreement with foreign governments. Some South American nations used this move as a precedent for claiming extended jurisdiction. Peru, in a Presidential Decree of August 1, 1947, claimed sovereignty over the continental shelf and a zone of 200 miles and soon thereafter began seizing U.S. tuna vessels in that area. This led to a major controversy with the United States which was followed by the Santiago Declaration of August 18, 1952, in which Peru joined Ecuador and Chile in a declaration of a 200-mile territorial sea. The political rationale was the concept that the distinctive coastal waters really belonged to the coastal countries in a common "bioma."

The great opportunity for Peru's fishery development came with the expansion of the world market for fish meal and oil. During the late 1940s the world production of fish meal increased to about 600,000 tons annually. About 20% of this came from nonoily fish or crustacean waste produced during processing operations and a few low value fish such as hake caught especially for fish meal. The balance came in small part from the waste of tuna, salmon, or herring processing but in major part from species of the herring and anchovy families caught especially for reduction to oil and meal.

Fish meal is sold as dried meal but also as solubles. The latter includes the water-soluble fractions obtained during pressing after cooking; if they are not dried and added to the meal, they are dewatered to a syrupy "stickwater" for transport and sale. The raw fish yield from 15 to 23% dried meal, depending on the efficiency of the equipment, and about 2.5% salable oil.

The market was expanding because the fish meal had been found to be an important additive to diets for poultry and young farm mammals. The fish proteins have some amino acids not available in grain that foster faster growth and healthier animals. The principal producing countries in the late 1940s were the United States (with about 40% of world production), Canada, and several countries of northwest Europe; about 80% of their production was consumed domestically. The market for fish oils was also expanding as a variety of food and industrial uses were discovered.

Peru's anchoveta resource was almost ideal for fish meal and oil production. It was very abundant close to shore where it was easy to catch. Depending on their size and season, the fish required no preservation on board the vessels if they could be unloaded in less than 24-48 hr. Thus simple vessels were adequate; the *bolicheras* (purse seine vessels) had to be equipped merely for handling the net and stowing the catch but required no extra power for towing a net such as a trawl. Fortuitously, nylon nets, which are much stronger and more rot-resistant than the cotton nets, became available in the mid-1950s. They reduced the cost and labor of cotton nets as well as providing economies of scale by making practical much larger nets and boats.

Limited fish meal production in Peru started in the late 1940s as an effort to use waste from bonito canning. The export market for bonito (an inferior kind of tuna) soon faded and the first fish meal plants turned to anchovy or herring. The first anchoveta plant was built in 1950. The total fish catches, almost entirely anchoveta, doubled between 1950 and 1955, and then exploded by a factor of 14 between 1955 and 1960 to 3.5 million tons.

They continued upward to 7 million tons in 1962, and to an alltime high of 12 million tons in 1970 (Fig. 6.2).

This expansion was a result of private initiative. The first plant was a joint venture between a Peruvian and a U.S. company. Other companies began operations, and by the late 1950s profits were reputed to range up to 50%. Credit was readily available, yards were constructed to build boats, engineering firms began to produce fishmeal processing equipment, and fishmeal companies financed boats in order to control their supply.[24]

Chimbote, the principal fishing port in the northern part of the country, had 14 fishmeal plants in 1958 and 31 by the end of 1962. Its population, which was estimated at about 17,000 in 1954, may have reached 85,000 by 1962.

The boat-building business surged. A survey of shipyards in 1963 counted 42 with capacity enough to more than double the number of registered boats in 1 year. The total hold capacity of the registered bolicheras was about 3000 t in 1954, 49,000 in 1960, and 162,000 by 1968.

By 1962, Peru surpassed Japan in total fish landings to become the nation with the largest fish catches in the world. The production, however, was 96% anchoveta, so the total value was much less than that of Japan.

The world markets for fish meal and oil did not expand proportionately with Peru's production. By 1962, Peru produced about 39% of the world supply and exported about 62% of the world exports. Prices fell drastically from a peak in 1959.

The price drop was a concern of fishmeal producers worldwide, who formed an international Fishmeal Exporters Organization in about 1960 to establish a system of export quotas. Peru became a member, but with its quota at 60% of the expanding market, the quota had little restraint on Peru.

Concern about the productivity of the anchoveta resource grew in the early 1960s. The Peruvian Government had had oceanic and fishery research scattered among the CAG, the Ministry of the Navy, and the Ministry of Agriculture. It attempted to coordinate this in the Consejo de Investigationes Hydrobiológicas in 1954. A UNDP/FAO project began in 1960 to collect accurate catch statistics and to study the dynamics of the fish resources.

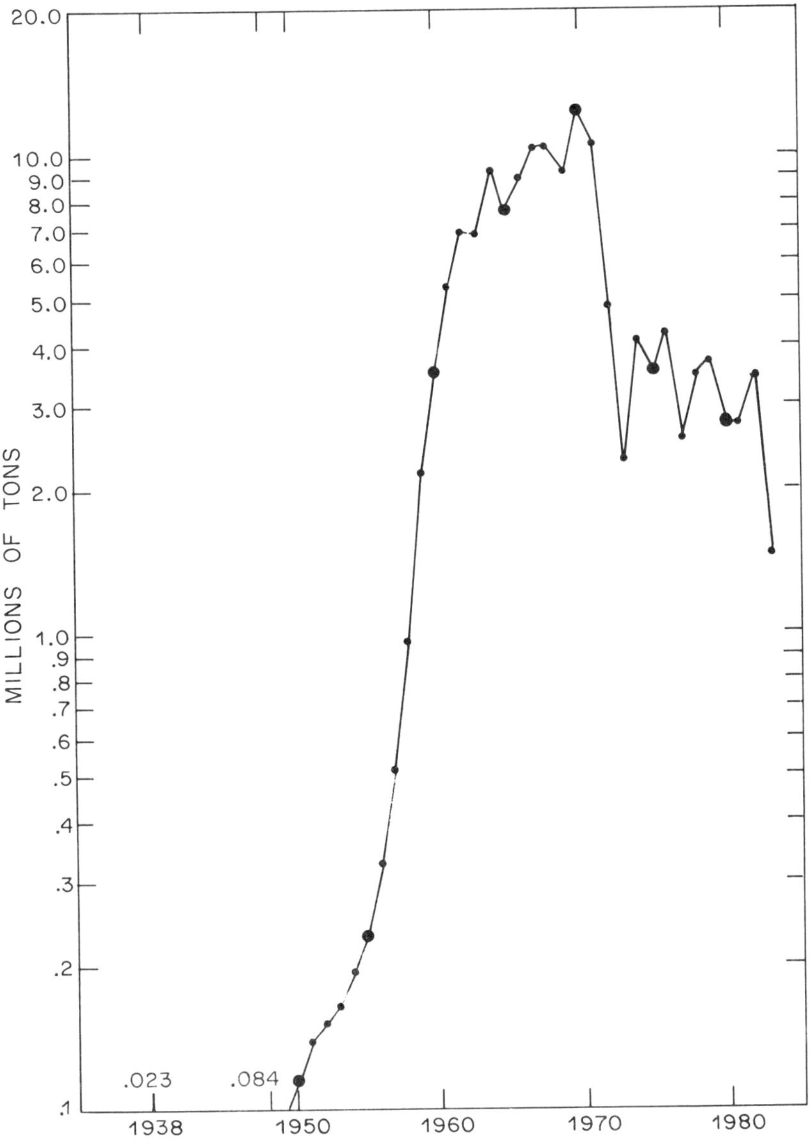

Fig. 6.2. Fish catches by Peru. Source: FAO "Yearbooks of fishery statistics, Catches and Landings," 1948-1983. (7)

This resulted in the scientific center, Instituto del Mar de Perú (IMARPE), which began in 1965 to make an annual review of the status of the resources in collaboration with foreign experts.

These reviews produced estimates of the sustainable yield for the fishery in the range of 7-10 million tons, but the estimates depended in part on the size of the guano bird populations. During the early 1960s the bird population was estimated as about 17 million, which consumed about 2.7 million tons of anchoveta. In 1965, however, a mild El Niño event occurred which starved the bird population down to about 4 million which would then consume only about 700,000 tons.[22]

Then in the summer of 1971-72, a full scale El Niño event occurred with a major intrusion of warm tropical water southward along much of the coast. The scientific studies revealed very low survival of young anchoveta from the September spawning, and further that the remaining anchoveta were unusually concentrated in pockets of cool water, where they were vulnerable to fishing. Management of the fishing for conservation became essential for the condition of the resource as well as critical for the future economic condition of the industry. Total fish catches for the country plunged from about 10.3 million tons in 1971 to about 4.8 million tons in 1972 and further to about 2.3 million tons in 1973.

This was not only a resource catastrophe but also an economic catastrophe for the country. Fish exports had grown to 27% of total exports by 1967; just ahead of copper at 26%,[24] and they were probably a greater proportion by 1971. Fishing for anchoveta was entirely closed from July 1972 to March 1973, which threw thousands of fishers and plant employees out of work.[25]

The Peruvian government reacted as it had to. It took over operation of the industry in May 1973 when the industry's total assets were estimated to be US$120 million with debts of US$227.5 million. This was done through creation of a state monopoly of boats and processing plants, the Empresa Pública para la Elaboración de Harina y Aceite de Pescado (PESCA-PERU).[26] The fleet of bolicheras and the number of fishmeal factories were both judged to be twice as large as would be needed even for a catch of 10 million tons.

Even before this action, while the fishmeal industry was at the peak of its expansion in 1970, the government had moved toward more national control. It had separated fisheries from agriculture in a new Ministerio de Pesquería in which it had placed IMARPE. In addition, it created within the Ministerio two parastatal bodies to execute the plans of the Ministerio in competition or collaboration with the private sector. These were the Empresa Pública de Servicios Pesqueros del Perú (EPSEP), in order to promote fish for direct human consumption, and the Empresa Pública de Commercialización de Harina y Aceite de Pescado (EPCHAP).[27]

The Ministerio prepared a five-year plan for 1971-1976 in which it projected a doubling or tripling of the food fish catch of 200,000 tons. It identified a hake resource capable of sustained yields of 100,000-300,000 tons. It began to negotiate a program of bilateral and multilateral aid. It gave special support to infrastructure for the artisanal fisheries and the fishery cooperatives. This included 16 coastal fish terminals, 12 inland fish distribution depots, help with construction of small vessels suitable for food fisheries (few anchoveta vessels could be converted), a fish inspection service, and consumer education programs.

During the late 1970s and early 1980s Peru still regarded the fisheries as vital national activities but it discovered the difficulties of developing their complex and changable resources. Unfortunately, the El Niño event of 1971-1972 was not the typical temporary phenomenon of earlier decades. Instead it apparently triggered a long-term change in the oceanic fish populations which had not recovered by 1983 (Fig. 6.2). This left the state monopoly on fishmeal production, PESCA-PERU, characterized as the most debt-ridden state fishing company in the world. After almost 10 years of increasing debt, the government took steps to reduce the wasteful bureaucracy and began to encourage joint ventures.[26,28]

The jack mackerel stocks flourished after the collapse of the anchoveta stock, and foreign fleets began to harvest up to several hundred thousand tons annually. The old anchoveta vessels were, however, mostly unsuitable for this fishery. The government had a general economic crisis, so it sought international assistance for development of a jack mackerel fishery.

A continuing high priority was given to development of the artisanal fisheries and programs to increase food fish consumption. EPSEP, which had been formed in 1970 to promote fish comsumption from local catches, was able to distribute only about 150 tons per month. Therefore the government arranged to purchase 1500-2900 tons per month from Soviet trawlers in late 1984. The government also continued to subsidize export of various food species.

Much more attention was also given to development of fish farming and freshwater fisheries. Both shellfish farming along the coast and trout farming in the mountain areas appeared to have potential, especially for export products. The upper Amazon basin in eastern Peru, which already supported a profitable export trade in ornamental fish, was surveyed for its potential food fish production.

Regardless of more than a decade of trouble and major government expenses to support the fisheries, they have continued to be regarded by the government as "a pillar of its development."[29] But the country's politics may change this status, because the plight of the fisheries was raised as an issue in the April 1985 elections.[28] The new president has appointed a new minister of fisheries who will, no doubt, be conducting more reviews of Peruvian fishery policies.

One analyst has characterized the events surrounding the Peruvian fisheries as myths that are frequently invalid. The international-level myth is that collapse of a fishery will have little effect on other fisheries except through the marketplace. On the contrary, Chile became involved in the anchoveta fishery in an effort to alleviate unemployment in its northern district. A national-level myth is that fishing is predominantly an economic activity. On the contrary, it is part of a political system within which it competes for attention and support. An individual-level myth is that the management of each stock is subject to an individual prescription. On the contrary, there are multiple objectives, related to other fisheries and changing over time.[30]

6.5 FAILED EXPECTATIONS--EASTERN CANADA

The immense fishery resources of the Grand Banks off Newfoundland were discovered by the earliest European explorers, and they have been fished ever since by European vessels as well as by the shore-based vessels and boats. The fishing and the settlements dependent on it spread, of course, to the grounds in the Gulf of St. Lawrence, the coast of Labrador, the coast of Greenland, the more southern coasts of Nova Scotia and New Brunswick, as well as the coasts of what was to become the United States.

The primary target of the fishing in the entire northwest Atlantic until about the middle of the twentieth century was the cod. They were caught mostly by line gear until the early twentieth century, after which much of the line gear was gradually replaced by trawl gear. In the offshore fishery the line gear was operated by fleets of schooners which dispatched dories to set and retrieve the gear. In the inshore fishery, dories could be launched from shore and be used either to operate small-scale line gear or to tend simple traps.

Most of the line gear catches were wet-salted in casks on board or salted ashore and dried. Either product was a staple food for the communities as well as an important item for shipment home to Europe or for sale abroad.

Good statistics on the total catches did not become available until after 1950, but during the 1920s and 1930s the Newfoundland people apparently caught 200,000-300,000 tons of cod annually; the people in the Maritime provinces and Quebec, another 75,000-150,000 tons.(31) During the same period the French fleets, although fluctuating greatly, caught as much as 175,000 tons in addition to landings on the French Islands of St. Pierre and Miquelon located just off the Newfoundland Coast. Further, the combined fleets of Denmark, Norway, Portugal, Spain, the United Kingdom, and the United States probably averaged at least another 100,000 tons annually off the coasts of the Maritime Provinces and Newfoundland. Thus, the cod catches alone probably fluctuated around a level of at least 0.5 million tons annually for many years.

During these same two decades, the markets for fresh and frozen fish began to expand. Other species of groundfish, including haddock, redfish, several species of flounders and hakes, as well as large quantities of the pelagic herring and mackerel, were targeted.

After World War II, fishing in the northwest Atlantic by European nations increased dramatically, much to the concern of Canada and the United States. They, therefore, negotiated the International Convention for the Northwest Atlantic Fisheries of February 8, 1949, for the purpose of conserving the resources and "...in order to make possible the maintenance of a maximum sustained yield..."(32) The countries signing included Canada, Denmark, France, Iceland, Italy, Newfoundland (which joined Canada later in 1949), Portugal, Spain, the United Kingdom, and the United States. Subsequently, Bulgaria, the German Democratic Republic, the Federal Republic of Germany, Japan, Romania, and the U.S.S.R. became parties. The Convention established the International Commission for the Northwest Atlantic Fisheries (ICNAF). Protocols later extended the responsibility of the Commission from just finfish to mollusks and to hood and harp seals. The convention area included the seas off the west coast of Greenland and the east coast of Labrador, south to and including the coast of Rhode Island in the United States.

The Commission headquarters were soon established in Halifax, Nova Scotia, and a staff began to assemble complete statistics on the fish catches by all countries in the Convention area. Canada and the United States expanded their research on all of the fish stocks and were assisted by researchers and research vessels from several European countries. Both statistical and research programs were later extended northward along the east coast of Baffin Island in Canada and southward along the U.S. coast to North Carolina.

The international cooperation in research became exemplary, and regular estimates were made of the sustainable yields of the many fish stocks under various fishing conditions. The first conservation measures were stipulations for minimum mesh sizes in the trawl nets.

Soon after Newfoundland joined Canada in 1949, the two governments established a Newfoundland Fisheries Development Committee. In a report in 1953, the Committee recommended government assistance to provide better boats and gear and better processing facilities in order to raise the productivity of the inshore fisheries. It paid scant attention to the offshore fisheries which were small and ill-equipped by international standards of that time.(33)

Both national and provincial governments provided modest amounts of funds to support the recommendations, but in 1962 another drive for assistance started. It continued to emphasize the inshore fishery, reflecting the sentiments of the great majority of fishers who preferred the status quo. Then in 1963, the Newfoundland government produced a monograph entitled *National Fisheries Development* which focused on the inshore fishery and the salt fish trade because, as it stated, the opportunities for profitable increases were probably the highest there.

Next came a report of the *Royal Commission on the Economic State and Prospects of Newfoundland and Labrador* in 1967 which recommended resettlement of people from the scattered inshore fishing villages to the offshore fishing and processing centers.

How then could the fishery of Newfoundland, making use of one of the best fishing grounds in the world while good fishery research was being conducted, become a social disaster? It was described in a report of a study commissioned in 1970 as follows:

> One hundred years ago the fishery provided the sole means of support for the Newfoundland economy. Today the fishery is probably the chief economic liability of the province - considering the great amounts of public expenditure that it absorbs against such meager returns... The problem of high levels of assistance combined with low returns appears to be concentrated in the inshore fishery pursued by outport fishermen.(34)

The traditional inshore fishery pursued mostly cod and lobsters from small "outports" scattered in the fjords along the coast. The total inshore catch declined from a level of about 215,000 tons in the mid-1950s to about 152,000 tons in 1970. But Newfoundland increased its offshore fishery (mostly large trawlers) from a catch level of about 45,000 tons in the 1950s to about 173,000 tons in 1970.

The extent of social change is better indicated when these figures are given on a more individual basis.(35) The inshore fishers, who numbered about 16,000 in the late 1950s, produced about 14 tons annually per person (on a full-time equivalent basis) with a value of about C$800. In 1970 they numbered about 14,000, and produced about 12 tons annually per person with a value of about C$1400; an increase in value scarcely more than the inflation.

The offshore fishers numbered only about 400 (also on a full-time equivalent basis) in the late 1950s, when they produced about 118 tons annually per person with a value of about C$6,000. In 1970 they numbered 1096 and produced 160 tons annually per person, with a value of about C$12,000.

During this period both the federal and provincial governments had increased their financial support of the industry. The net total government assistance averaged about C$10 million annually in the late 1950s and expanded to nearly C$30 million annually in the late 1960s. This was equivalent to an average of about 80% of the total value of fish landed in the late 1950s; 134% in 1968.(36)

The government assistance program resulted in no demonstrable improvement in the inshore fishery, whereas the offshore fishery with little public support, after lagging in the 1950s, expanded largely with private initiative in the 1960s. The problems of the isolated outport fishing communities increased because of the limited local fishery resources and the virtual absence of other employment opportunities. The businesses had no incentive to invest in fishing or other activities. The people had to move to larger communities and change their artisanal life if they were to improve their incomes.(37)

The problems of the Newfoundland fisheries were more extreme than those of the other Atlantic provinces which had a much smaller proportion of outport communities. The total landings in Newfoundland were about the same size in the early 1950s as the total of the other Atlantic provinces, which were less dependent on cod. The other provinces, especially Nova Scotia, with a larger proportion of offshore fishing, increased their share of Canadian Atlantic catches to about two-thirds in the early 1970s.

The promise of the offshore fishing to Canada was not realized, however, because of the intense foreign fishery (Fig. 6.3). The Canadian catches remained at about half of the total catches by all nations until the peak of about 2.6 million tons in 1968. Then came a decline to a total of about 2.0 million tons in 1975, with an even smaller Canadian share of about 0.85 million tons.[38] Moreover, the declines in value were even greater because offshore fisheries had turned to less-valuable species as the favorite market species became depleted.

Part of this new fishery was for pelagic species. Herring, mackerel, and capelin catches each increased by hundreds of thousands of tons annually, then declined precipitously, apparently due to natural causes.

The fishing restraints under ICNAF had been merely minimum mesh sizes on trawl nets until 1970, when catch quotas for haddock were agreed upon. This practice was expanded until all groundfish stocks came under a quota system in 1974. In addition, the total allowable catches (TACs) were computed on a more conservative basis than maximum sustainable yield (MSY).

Only a modest international inspection system was developed by ICNAF, so enforcement and statistical data were largely the responsibility of national efforts which "were frequently less than adequate."[39]

By 1975, the groundfish system was considered to be in a crisis.[40] The weaknesses of the offshore system were (1) an inability to catch more fish; (2) the high cost of acquiring and operating fleets, especially because of the oil prices; (3) practices mainly restricted to producing frozen products; and (4) the quality of plant management. The inshore systems were in worse condition because of (1) overcapacity for the size of

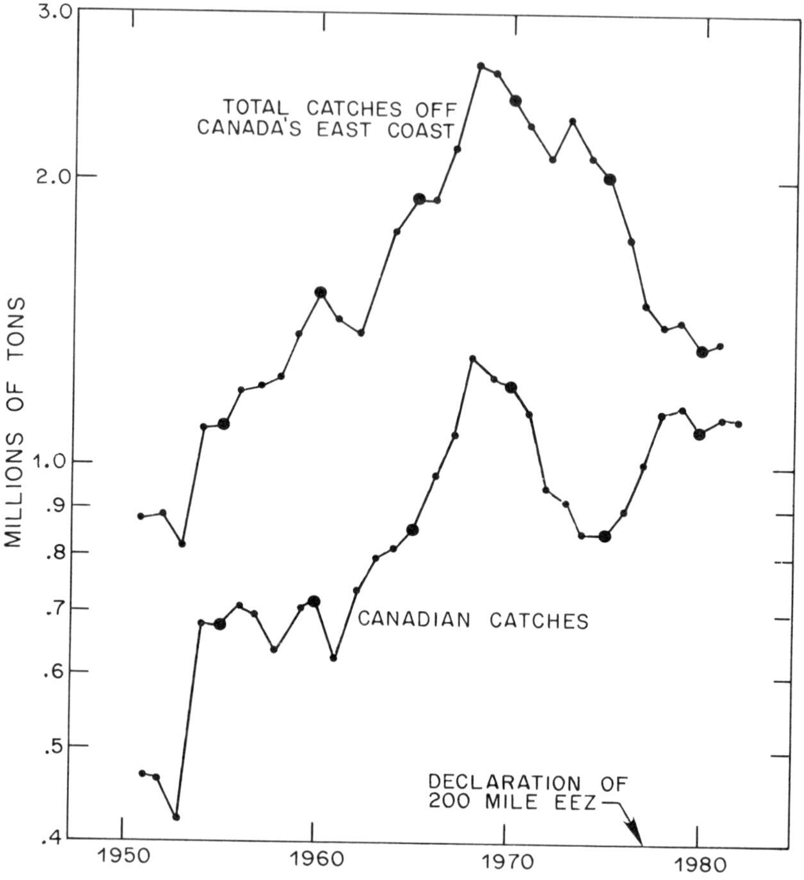

Fig. 6.3. Fish catches off the eastern coast of Canada. Source: Statistical Bulletins of ICNAF.(38)

the resource, (2) small-scale processing operations, and (3) limited financial capability.

The obvious condition of the stocks and the difficulties of enforcement had led Canada to favor extended jurisdiction over its coastal fisheries instead of reliance on ICNAF. It declared in 1971 that the Gulf of St. Lawrence and the Bay of Fundy were Canadian fishing zones and negotiated a gradual phaseout of the foreign fishing there. Then Canada joined with many other coastal nations in 1974 to support a 200-mile EEZ at the Caracas meeting of the Third Law of the Sea Conference. This consensus led to national action, and all four northwest Atlantic coastal nations, Denmark, Canada, France, and the United States, declared 200-mile EEZs in early 1977.

Thus Canada almost established national control over the Atlantic fish stocks off its coasts. Two problems remained. France had negotiated special concessions to continue fishing in the Gulf of St. Lawrence, and had widely overlapping 200-mile zones with Canada offshore from its islands of St. Pierre and Miquelon just south of Newfoundland. The jurisdiction of the two countries had not been settled by mid-1985, but fishing was continuing under temporary arrangements.(39) In addition, the southeastern tip of the Grand Banks is outside 200 miles, as is Flemish Cap, a shoal east of the Grand Banks. Both of these are important cod spawning and fishing grounds.

With the broad acceptance of a 200-mile EEZ, Canada gained leverage for negotiation with foreign countries. It allowed access to some little used stocks and made arrangements to overcome weaknesses in the fishing industry. The federal government attempted to assist the offshore industry by developmental charters that were designed to encourage production and processing of non-traditional species.(41) As many as 50 arrangements were made to use foreign vessels in directed fisheries. The companies, most of which were large, gained experience in processing and marketing, but the net benefits appear to have been marginal. Other sales were made at sea from Canadian fishing vessels "over the side" to foreign processing vessels. These, however, left the fishing sector at odds with the processing sector on shore over a perceived loss of jobs and marketing opportunities.

A review of the cost-effectiveness of the Newfoundland groundfishery, based primarily on data from 1978, found that the entire fishing sector was heavily subsidized and that different kinds of vessels varied widely in economic performance.(42) Surprisingly, some classes of large vessels were much less cost-effective than smaller vessels, and also the socioeconomic loss was about the same in both inshore and offshore fisheries.

The review predicted that an immediate and complete removal of the subsidies would be a major shock to the entire industry.

The situation grew worse, and in January 1982, a government Task Force on Atlantic Fisheries was established because of the threatened financial collapse of five major fish companies: two in Nova Scotia, two in Newfoundland, and one in New Brunswick.(43) In its report, the Task Force encouraged concentration and restructuring of the corporate sector, much of which owned large trawlers as well as the shoreside processing plants. This was strongly criticized as an assault on the inshore fishers and the more efficient independent vessel operators.(44)

By 1985, Canada had made some progress in restricting entry into its fisheries through limited licensing, but it had been unable to deal with the weakness in port markets due to the large number of landing points, the weakness in end markets due to the structure of the processing industry versus that of the world competition, and the regional underdevelopment that was determined in large part by a settlement pattern based on fishing opportunities of previous centuries.(39)

An examination of the fishery decision-making process in several of the Atlantic fisheries resulted in a castigation of the government procedures. It pointed to the vulnerability of small fishing communities to reactive government decisions and to the power of the large corporate fish processing firms. It suggested that a much more humanistic approach was essential--one that placed community interests ahead of private interests.(45)

Furthermore, Canada may encounter additional tariff restraints in its principal foreign market. Canada was the world's leading nation in fish exports during the early 1980s (valued at about US$1.3 billion annually) when its principal customer was the United States. Most of its exports came from the Atlantic Provinces and competed directly with products from the same species of fish caught off New England.

A New England association of fishers, fishers' cooperatives, and processors involved with catching and marketing fresh Atlantic groundfish (whole or filleted) sought additional U.S. tariffs because the producers or exporters of such products in Eastern Canada received government assistance. In a petition filed August 8, 1985, the association alleged that benefits to the fisheries from Federal and Provincial programs constituted subsidies under U.S. law that entitled U.S. fisheries to tariff protection.(46) The petition listed the following Canadian Federal Programs:

Fishing Vessel Assistance Program
Fisheries Improvement Loans Program
Special Recovery Capital Projects Program
Fishing Vessel Insurance Plan
Import Duty Remission
Unemployment Insurance Act of 1971
Industrial and Regional Development Program
Investment Tax Credit
Government Equity Infusions
Program for Export Market Development
Development of Fisheries and Oceans Marketing Services
Assistance for the Construction of Icemaking and Fish Chilling Facilities
Atlantic Fishery Management Program
Federal Assistance for Bait.

Joint Federal-Provincial Programs included the following:

Economic and Regional Development Agreements Program
Newfoundland and Labrador Development Corporation.

Also listed were the programs supported by all of the Atlantic Provinces of which those from Newfoundland were the most extensive. The Newfoundland list included:

Fishing Ships Bounty Program
Fishing Vessel Assistance Program
Vessel Rebuilding Grant Program
Loans and Loan Guarantees from the Newfoundland Fisheries Loan Board
Newfoundland Bait Services Program

Sales Tax Exemption for Fishermen
Loan Deficiency Guarantee Program
Secondary Processing Interest Subsidy Program
Rural Development Loan Program
Production Machinery and Processing Technology Program
Market Development Information.

After a hearing on August 28, 1985, the U,S. International Trade Commission held that there was a reasonable indication that an industry in the United States was threatened with material injury by reason of imports of fresh Atlantic groundfish, whole or fillets, which were allegedly subsidized by the Government of Canada.

Then the U.S. Department of Commerce determined on January 2, 1986, that the Canadian exports to the United States were subsidized by 19 Canadian Federal and 20 Provincial Government programs. The net value of the subsidies was at first estimated at 6.85% of the value of the Canadian product, and then in March 1986, it was determined to be 5.82%, an amount that U.S. importers would be required to pay. The final decision of the U.S. International Trade Commission on April 29 1986, agreed with respect only to fresh whole fish, but it rejected imposition of any tariff on fillets. The controversy seemed likely to continue.(47)

6.6 CONTROLLED DEVELOPMENT--JAPAN

Almost everything about the Japanese fisheries is superlative. The country's population, currently about 120 million, crowded on islands smaller in area than the state of California, has always been heavily dependent on the sea. In most of the years of this century Japan has had the largest catch of any nation in the world. At least three-fourths of this is caught currently within its 200 mile zone, but its fleets fish around the world on the high seas or within the 200-mile zones of some 16 countries with which it has agreements. In recent years it also has been the world's largest importer of fish and, of course, has been the world's largest market for fish. Its annual fish supply is about 125 kg/person (round weight), and, even so, a large proportion of its people would increase their use of fish because they like it and because they believe that it is good for the health of their families. Its fishery products are characteristically of excellent quality, and for the past 20 years, fish prices have increased at a much faster rate than meat prices.

All of this is done through popularly supported government and cooperative organizations that both manage the resources and develop the fisheries. These comprise a system that has been firmly established in law and custom for many years despite profound changes in almost all operational aspects of the fisheries.

Japan's annual fish production has increased almost steadily throughout this century. It reached 1 million tons in 1905, surpassed 2 million tons in the early 1920s, and 4 million tons in the mid-1930s.[48] Its increases were interrupted by World War II, but production regained prewar levels in the early 1950s, and exceeded 10 million tons in the early 1970s. Further increases have been slow because of the rise in oil prices and the introduction of 200 mile EEZs in the late 1970s, but there were about 12 million tons in 1983 if fish farming is included (Fig. 6.4). These figures do not reveal, however, the major changes in the contribution of different fisheries and different species during recent decades, which need to be elaborated if the diversity and resilience of the system is to be realized.

Japan divides its fisheries into three sectors: (1) coastal fisheries pursued with boats or vessels less than about 10 gross tons, using lines, gill nets, traps, and set nets; (2) offshore fisheries operating vessels more than 10 gross tons, mostly within 200 miles of Japanese ports and using trawls, seines, drift nets, and pole and line; and (3) deep-sea (distant-water) fisheries operating mostly beyond 200 miles and employing gear similar to that of the offshore fisheries plus longlines.

Production from the coastal fisheries has been almost steady since 1955 near a level of 2 million tons. Thus their contribution to total production has declined from

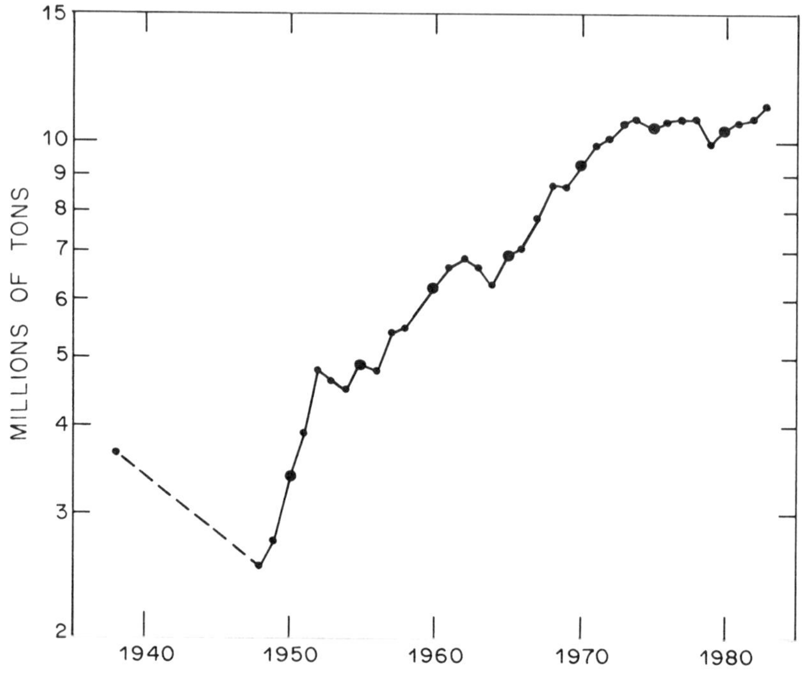

Fig. 6.4. Fish catches by Japan. Source: FAO "Yearbooks of Fishery Statistics, Catches and Landings," 1948-1983.(7)

about 40 to 20% in the three decades. Production from the offshore fisheries increased rapidly, however, from about 2 million tons in 1955 to over 6 million tons in the early 1980s, assisted in the latter years by a major recovery of the sardine populations near Japan. The deep sea fisheries followed a very different course, increasing from about 0.8 million tons in 1955 to a peak of about 4 million tons in 1973, followed by an abrupt decline to only about 2 million tons in 1980 and no recovery since.(48,49) Fish farming, of which about 90% is marine, and 10% freshwater, has been steadily increasing to reach a production of about 1.3 million tons in 1983.

All of these fisheries have experienced major problems in recent decades which in other countries might have led to social disasters, but which in Japan have been ameliorated, if not solved, by its unique political and administrative system. The basis of this system is the long experience with and acceptance of private rights in fishing. The fish resources have been removed from the commons and placed under a system of democratic control in which every fisher can have a voice and yet prefectural and national concerns can be accommodated.

The recent system has evolved from that of the feudal governments of the nineteenth century, in which the shoguns controlled the waters, and allowed fishing to fishers who paid taxes. After the end of the feudal regime, the Meiji Government attempted to nationalize the waters but after a few years they recognized the traditional rights of the fishers. The Fisheries Law of 1901 provided for fishing rights, scarcely changed since the feudal regime, plus provisions for fishery regulation and fishery associations. Then, in 1910 the law was clarified to define several kinds of rights and to give fishing rights the same legal status as property rights. Further changes were made soon after World War II, especially to grant fishing licenses only to persons or enterprises actually engaged in fishing, to entrust administration of fishing rights to fishery cooperatives or similar organizations, and to establish fishery coordinating committees for each sea area.(50)

The complexity of the fisheries devolves into the laws which, according to some, may defy a rational explanation. But numerous kinds of fishing rights are granted to cooperatives or to enterprises for the varied types of fishing. Each is an exclusive right to carry out a specific operation in a "demarcated area" along the coast or in a "designated fishery" of an ocean region. Apparently all kinds of fishing rights are subject to debate and approval by either cooperative organizations or coordinating committees. They are temporary, being granted usually for a period of 5 years.

The cooperatives were established originally to control the fishing rights in the small-scale fisheries of coastal and inland waters. These now constitute the area cooperatives to which every fisher of the coastal areas must belong. In addition, a few cooperatives have been established for participation in the offshore or deep-sea fisheries such as tuna longlining or trawl fishing. The nonco-op fishing enterprises, however, dominate the offshore and deep-sea fishing. They include several of the largest fishing enterprises in the world, three of which were listed in 1986 by *Fortune* magazine (August 4) among the 500 largest industrial corporations in the world outside the United States--Taiyo Fishing, no. 104, Nippon Suisan, no. 225, and Nichiro Gyogyo, no. 417.

In the offshore and deep-sea fisheries there are prefectural or national concerns, so these fisheries are licensed by prefectural governors or by the national Minister of Agriculture, who is responsible for the fisheries. Issuance of these licenses for "designated fish eries" is controlled to promote conservation, discourage overinvestment, allocate fishing among enterprises, prevent excessive competition, and minimize conflicts among enterprises.(51)

The fishery cooperatives of Japan are organized into national federations, the Zengyoren; prefectural federations, the Kengyoren; and local associations, the Gyokyo. The Gyokyo totalled 3072 in early 1985 with more than 0.5 million members. They operate savings and credit services, fuel and other supply services, cold stores, and various advisory and educational activities, and they market at wholesale level much of the fish sold in Japan. They work closely with related co-ops such as fish processor and credit co-ops. Their total annual business activity in recent years has been on the order of ¥170,000 million or about US$700 million.(52)

The Zengyoren have an active fishery policy department. They participate in international negotiations and take positions on numerous domestic issues such as coastal pollution. The Kengyoren, in a similar fashion, work closely with prefectural and municipal governments on fishery issues.

The teamwork of the cooperatives and government agencies (perhaps the cooperatives should really be considered as local government agencies) has managed profound changes in Japanese fisheries since the 1940s. The rather steady increase in total production (Fig. 6.4) masks the social and economic stresses that have occurred due to technical development and changes in markets, resources, and international agreements.

The technical development befits Japan's status as a major industrialized nation. It can produce, or readily obtain through its immense trading firms, the world's best fishing vessels, equipment, gear, and processing facilities. Its fishing equipment and facilities have been steadily upgraded. In its technology and its business organization, it has become the most proficient fishing nation in the world.

Such technical changes have been followed by social disasters in the small-scale fisheries of many countries but not in Japan, where social stresses certainly occurred but were managed. These fisheries employ 75-80% of the fishery workers in Japan, although employment in them *decreased* by about 40% from the 1950s through the 1970s. During these three decades the production of the coastal fishing remained a little below 2 million tons, but the production from fish farming increased from from less than 0.2 million tons to about 1 million tons. The boats used by the small-scale fisheries changed from about 99% without power in 1955 to 97% powered in 1978, along with a steady increase in their average size. These fisheries were entirely under the control of cooperatives, but the government assisted substantially with projects to reduce pollution; improve the environment with artificial reefs; stock young fish; and modernize refrigeration, market, and port facilities. In addition, the government promoted more consumption of species such as sardines and mackerel as they have become abundant. This has included advertising, new recipes, new product research, and better quality control. A result has been a rate of increase in household income of the small-scale fishers substantially better than that of urban workers, although slightly smaller than that of agricultural households.(53)

Changes in the offshore and deep-sea fisheries were much more turbulent, especially during the 1970s. Japan had expanded its offshore fisheries in the northwest Pacific and its deep-sea fisheries around the world in the 1950s and 1960s. It was heavily engaged in all of the fisheries between Alaska and Siberia in the Bering Sea and North Pacific Oceans.

Major changes occurred, moreover, in its domestic fishery for sardines. The production had been more than 1 million tons annually during the 1930s and 1950s, but natural causes led to a collapse to almost none in the 1960s. Then there was a resurgence during the 1970s and 1980s when production increased from 0.7 million tons in 1973 to 3.7 million tons in 1983.

Many more changes occurred in the 1970s as international pressure for extended fisheries jurisdiction increased and as it was declared by many countries in about 1977. Production of Alaskan pollock rose to a peak of about 3 million tons in the early 1970s, then declined to half that in 1980. Pollock had been a large proportion of the catches of north Pacific deep-sea trawlers, which declined from 1 million tons in 1972 to 0.2 million tons in 1978, and of the catches of the West Kamchatka trawl mothership fleet which declined from about 1.3 million tons to about 0.5 million tons. Production of Pacific mackerel declined from 1.4 million tons in 1979 to half that in 1982. Production of the crab gillnet fleet was halved in the early 1970s. Production of the north Pacific salmon gillnet fleet was more than halved during the 1970s.(54)

The Japanese government reshuffled its fishing fleets through its control of licensing. More than 1500 licenses were nullified during the 1970s, and only 135 were added to a few fleets fishing on resources which could produce more.(55)

The economic impact of these reductions in the resources available to Japanese fishers was further aggravated by the increase in fuel prices during the early 1970s as well as similar rates of increase in the costs of ropes and gear.

The Japanese government undertook many steps to assist the fisheries but went far beyond simple increases in subsidies. It especially protected the coastal fisheries and encouraged their transition from fishing to fish farming. It promoted consumption of species in large supply, such as sardines and mackerel, and adoption of stricter quality control in all fishery products. It extended cooperation to foreign nations as part of fishing agreements; including grants in aid reaching ¥12 billion (about US$50 million), training vessels, shore facilities, technical advice, and education of trainees. It encouraged a private organization, The Overseas Fishery Cooperation Foundation, to support foreign assistance as a part of its strategy to increase or maintain access to EEZs. It provided over ¥150 billion (about US$650 million) to fishers who lost jobs in the vessel reduction program and to finance continuing operations by the remaining vessels. This was accompanied by a mutual assistance program in which the employed fishers paid increased license fees in order to assist those who had lost jobs. It reduced fuel consumption in the fisheries by mandatory reduction in engine speeds and by incentives to construct more efficient vessels.(48)

Unquestionably, the difficulties of the Japanese fisheries have not been eliminated, nor are they likely to be. The catches of Japanese fishers near Japan can increase only through fortuitous resource increases or gradually and modestly through their program of environmental improvement. They seem far more likely to decrease with a failure of the sardine resource, as has happened previously off Japan, as well as with similar species off California, Peru, and Northwest Europe.

Their catches off other countries can be continued only through bargaining with each country, and all countries see the fish of their EEZs as resources to be used in their domestic interest.

The only significant possibilities for increased fish supply are through fish farming or increased imports. Japan is already heavily involved in fish farming and can increase it substantially (see Chapter 7). Japan can also increase its imports through its own large fishing companies, many of which have transnational affiliates. A recent example is importation of salmon from Canada and the United States, which grew from less than 10,000 tons in 1974 to more than 100,000 tons annually in 1982 and 1983. The major source is Alaska, and this is arranged, no doubt, largely by Japanese affiliates which are informally estimated to control about half of the investment in Alaskan salmon processing firms.

6.7 TROUBLE AHEAD--UNITED STATES

The 200-mile zone of the United States produced about 7 million tons annually or about 9% of the world's commercial fish production during the early 1970s, more than half of which was caught by foreign vessels. In those years, the United States was also importing about 1 million tons of fish products which, in round weight equivalents, was nearly as much as the foreign fleets were catching. In the same years the United States exported only about 100,000 tons (product weight) annually.

This time in the early 1970s marked the end of a relatively stable period in the U.S. fisheries. The U.S. catches had increased little since the 1940s, whereas imports had increased steadily from a level of about 200,000 tons in 1945 to about 850,000 tons in 1970 - a growth rate of more than 5% annually (Fig. 6.5). U.S. exports had approximately doubled during the 1960s, but were primarily a few specialty products and never had a total value greater than 15% of the value of imports.

The U.S. people, now numbering about 235 million, eat relatively little fish. The annual per person consumption (edible weight) of commercially caught fish has almost stagnated at about 5 kg since 1909, the earliest year of records.(56) It declined to about 4 kg during the depression years of the early 1930s and the war years of the early 1940s. It has increased to about 6 kg in 1983 and 1984. Prior to this increase, the additional supply of edible fish necessary to keep up with population increases came from the imports. The total annual per person supply (round weight basis, including imports and industrial products) was about 20 kg during the 1950s, increased to about 29 kg in the late 1960s, but decreased to about 24 kg since the early 1970s.

A unique feature of the U.S. fish supply, however, is the considerable catch by recreational anglers who number nearly one-fourth of the total population. Their recent catches have probably added about 1.5 kg (edible weight per person) to the annual commercial supply.(57) But a total of about 7.5 kg per capita is still only about one-fourth the consumption of the Japanese.

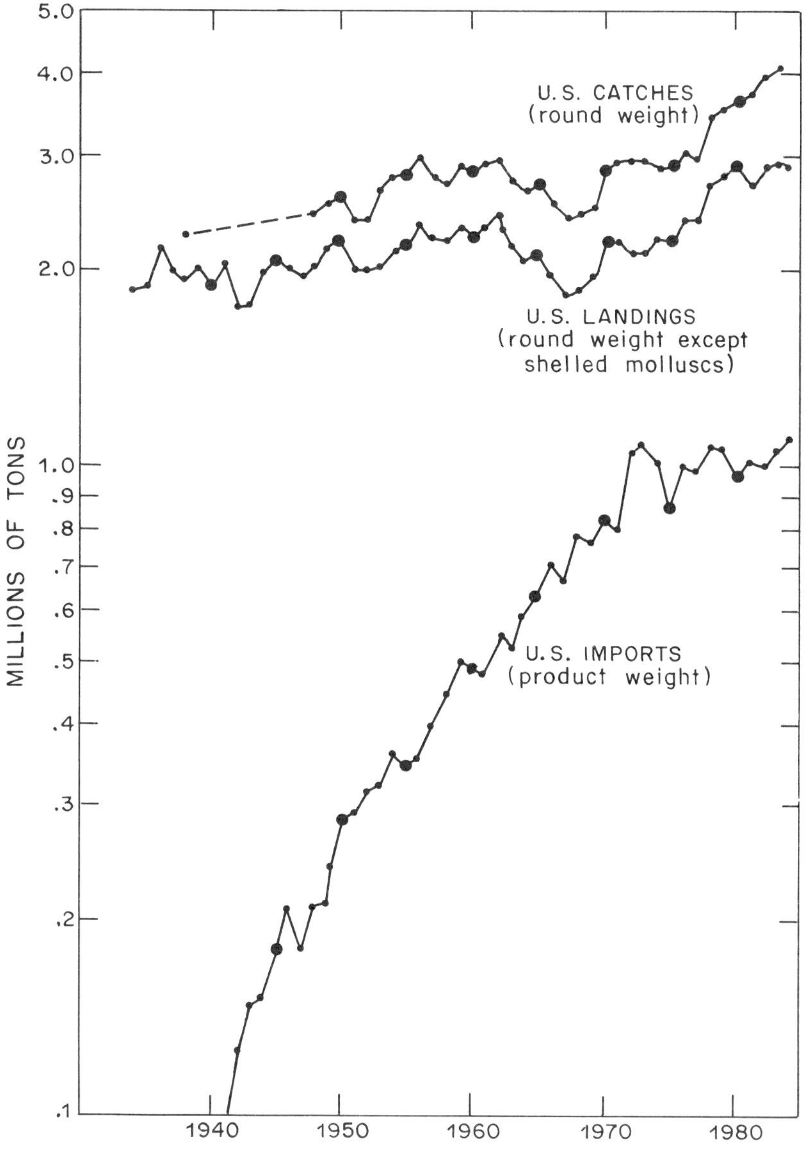

Fig. 6.5. Fish catches, landings, and imports by the United States. Sources: FAO, "Yearbooks of Fishery Statistics, Catches and Landings," 1948-1983,(7) and U.S. Annual Reports, "Fishery Statistics of the United States."(56)

The U.S. fishery policy has long been dominated by the concept of fish as a resource of the commons and by a strong public concern for conservation of the environment and natural resources, including the fish. Most of the charters granted by England to the 13 original colonies gave the settlers the right to fish in common waters. During the nineteenth century, the fish stocks of most of the fresh waters became depleted and most stocks of fish were reserved for personal use through prohibition of gear suitable for commercial use (see Chapter 5). The priority for personal (now primarily recreational) use has been extended recently to several salt-water species popular with anglers.

By the 1930s angling had become a popular recreation and it has increased in popularity since. The numbers of anglers during the 1960s and 1970s increased at a rate about double that of the population increase. Now the number of anglers is more than 300 for every commercial fisher. Thus, the political perceptions of the use of the fishery resources have been heavily influenced by those of the anglers. The concept of the public right to use the resources is a cornerstone of the fishery resource management programs.

The fishery management prior to 1950 was almost entirely a responsibility of the states; neither the states nor the federal government exercised any significant regulation of commercial fishing outside 3 miles, the limit of their jurisdiction. Thus the fishery programs were primarily to manage the resources for anglers in fresh and estuarine waters. The management included an extensive program of stocking young fish and restraining fishing in order to conserve the stocks and spread the catches among all who wanted to fish.

The management programs for recreational fisheries accelerated in the late 1940s as concerns arose over the conservation of more and more stocks. Congress passed a landmark bill to assist the states in management of the recreational fisheries. It was the Federal Aid in Sportfish Restoration Act of 1950 (Dingell-Johnson Act) which earmarked funds from a federal exise tax on fishing tackle for research and management programs by the states, primarily in fresh waters.

The federal government acted also to conserve the commercial stocks through negotiation or renegotiation of several fishery conventions. These included the International Whaling Convention of 1946, the Northwest Atlantic Fisheries Convention of 1949, the Inter-American Tropical Tuna Convention of 1950, The International North Pacific Fisheries Convention of 1952, The Great Lakes Fisheries Convention of 1954, and eventually others. It also refurbished the Halibut and Salmon Conventions with Canada, which had originally been negotiated in the 1920s and 1930s respectively (see Section 5.3).

These conventions required a strengthened federal role in fishery management, which had been largely restricted to research and collection of statistics (except in the territory of Alaska prior to statehood in 1959 where regulation had been a federal responsibility). Congress therefore passed the Fish and Wildlife Act of 1956, which focused attention on commercial fisheries within the Fish and Wildlife Service by creation of a new Bureau of Commercial Fisheries.(58) Further, because the marine fishery management by the states was so badly coordinated, Congress authorized interstate compacts among neighboring states.

All of these actions had conservation as the primary objective. Development may have been mentioned but was not emphasized in any of them.

The complexities and limitations of the conservation programs emerged during the 1960s when the world fish catches increased by an amount greater than in any other decade in history. The catches rose steadily from 40 million tons in 1960 to about 67 million tons in 1970 (Fig. 3.1), while U.S. catches experienced a decreasing trend, although both 1960 and 1970 catches were about 2.8 million tons (Fig. 6.5). A significant part of the increase in world catches came from European vessels fishing off the Atlantic Coast of North America and from Asian and U.S.S.R. vessels fishing off the Pacific Coast, especially off Alaska. This fishing was subject to some international control under treaties, but the controls had clearly failed to prevent depletion of numerous stocks.

In addition to doubts about conservation, it became abundantly clear to U.S. fishers that the United States was not using many of the fish resources off its coasts, whereas it was importing more than 1 million tons of fish products annually. In fact, its fish imports in the early 1970s were greater than those of any other nation in the world.

Congress responded to the increasing concern about ocean resources with a Reorganization Plan No. 4 of 1970 (84 Stat. 2090), which established the National Oceanic and Atmospheric Administration (NOAA) in the Department of Commerce. This created in NOAA the National Marine Fisheries Service (NMFS) to which was transferred the duties of the Bureau of Commercial Fisheries from the Fish and Wildlife Service of the Department of Interior. Also included in NOAA was the National Sea Grant College Program from the National Science Foundation. The transfer to the Department of Commerce was intended to provide stronger support for the business of fishing, for marine fishery conservation, for reduction of the foreign fishing, and for development of U.S. fishing to reduce the volume of imports.

The United States soon joined in a United Nations resolution of December 18, 1970 in support of an equitable international regime for the conservation of the living resources of the sea and of preferential rights in the resources for coastal nations. This evolved into a major step toward a new LOS with respect to fisheries through the consensus that emerged from the Caracas meeting in 1974 of the Third United Nations Conference on the Law of the Sea (see Section 4.1). A large majority of the countries attending, including the United States, supported a 200-mile EEZ. The United States enacted Public Law 94-265, the Fishery Conservation and Management Act of 1976 (FCMA), which was substantially consistent with the U.N. action, and which became effective March 1, 1977. Canada and numerous other countries enacted similar legislation at about the same time.

The Act, as its title indicated, had the primary purpose of conservation and the management action necessary to achieve it. It did include as its sixth and last purpose "to encourage the development of fisheries which are currently underutilized or not utilized by United States fishers, including bottom fish off Alaska." Its declaration of policy, however, did not mention fishery development.

But a fishery development policy was implied in the FCMA, and national efforts to increase assistance soon began. The United States dusted off an Act of July 1, 1954, that had become popularly known as the Saltonstall-Kennedy (S-K) Act. It was actually an amendment of some fishery assistance acts of the 1930s which had provided for purchase of surplus fishery commodities. The S-K Act designated up to 30% of the customs receipts on fishery products for the purpose of carrying out fishery research and development projects. This program had contributed modestly to fishery research for more than two decades, but in fiscal year 1979 when US$8.2 million was appropriated, it was chosen as the primary vehicle to assist the industry in efforts to take advantage of its priority rights in the 200-mile zone. The amounts appropriated annually soon rose to more than US$30 million during the early 1980s.

This then became the main federal effort to assist with development of the U.S. fishing industry, although the amount of money was slight when compared with the total value of U.S. commercial landings (US$2.4 billion with a product value of US$5.2 billion in 1984), the total value of U.S. fish imports (US$5.9 billion in 1984), or the total value of U.S. fish exports (US$900 million in 1984). It was, moreover, an addition to federal assistance to commercial fishing that had been provided since the 1930s under numerous acts of Congress. This included research on product technology, provision of market statistics, promotion of product standards, a fishing vessel construction differential study, and fishing vessel mortgage insurance.[58] It also included steps to coordinate the numerous other federal assistance programs that could be available to the fisheries.

An immediate step under the new emphasis was to contract for a study of the market opportunities and for a development plan. The market report [59] reviewed the resource availability and estimated that the U.S. catches could be more than doubled. It identified some very large export markets for species underutilized by the United States, identified numerous problems, and suggested a major expansion of export market news gathering systems. It suggested, however, that the major opportunities lay in replacement of the imports by domestic products. This was supported by observations that there was a steady increase in gross demand for seafood, especially for frozen product forms most of which were imported, that imports could be replaced with a lesser risk for production investments, that prices for fish had been increasing faster than prices for meat products, that people were recognizing the nutritional benefits of eating more fish, and that the current market for groundfish products was already larger than US$500 million annually and growing.

The development plan[60] evaluated in more or less detail eight potential fisheries for underdeveloped fish resources, which were selected from the many other possibilities. The eight were Alaskan groundfish, Pacific hake, Gulf of Mexico groundfish, Atlantic whiting, Atlantic mackerel, Atlantic squids, Pacific mackerels, and Hawaiian Island fisheries. It was a very complex projection which considered the increased employment, investment spending, retail sales in the United States, and the impact on the balance of trade.

The report concluded that the underutilized resources could be developed by 1990 with substantial benefits to U.S. society. It was based on determinations that, (1) the net present value in 1979 of such development was over US$1 billion, (2) over 43,000 jobs would be created, (3) an active governmental role in planning and public investments would enhance the benefits, (4) the U.S. trade balance would be improved by over US$1.5 billion annually, (5) markets for seafood products were strong and would continue to improve, (6) the promise of long-term return would be enough to encourage private investment, and (7) impediments to development existed in areas where federal action was appropriate.

The report was followed within days by a Final Report of the Department of Commerce Task Force on Fisheries Development which proposed a policy and program.[61] The key elements were (1) a cooperative federal government-industry program; (2) effective cooperation among industry and state, local, and federal governments; (3) consideration of an extension of a tax deferral program, already available to fishing vessels, to shoreside processing facilities; (4) negotiations to improve access to foreign markets; and (5) a broad review of impacts and benefits of regulations plus efforts to streamline the fishery management process.

Congress passed the American Fisheries Promotion Act (AFPA) of 1980,[62] which was a series of amendments to previous acts. In Part A, the S-K Act was changed to strengthen the involvement of and benefits to the fishing industry and to appoint at least six foreign service officers to collect information on fisheries in overseas embassies. Part B extended the Merchant Marine Act to provide loan funds that could be used by fishing vessel or facility owners to avoid default on private loans. Part C strengthened the FCMA in favor of domestic fishers and required that each foreign fishing vessel in U.S. waters carry a U.S. observer. Part D strengthened the Fishermen's Protective Act of 1967, which protected fishers against loss of vessels or gear due to foreign fishing.

The Department of Commerce followed through with a new focus for the Saltonstall-Kennedy Program.[63] Establishment of regional fisheries development foundations as nonprofit entities representing the fishing industry was encouraged in the Alaska, Great Lakes, Gulf and South Atlantic, mid-Atlantic, New England, Pacific, and west coast areas. These participated in an S-K Grants Program along with industry, state and local governments, and universities which provided a total of about US$8 million to US$14 million annually from S-K funds. In addition from US$13 million to US$26 million was included in the operations of the National Oceanic and Atmospheric Administration for related activities. The projects included in the numerous programs involved a broad assortment of activities such as trials of modified fishing gear, vessel safety, training for fish retailers and restaurant buyers, improvement of seafood quality, market assessments, and economic studies.

Statistics are now available to assess overall changes in the fisheries between 1979 and 1984.[56] The results are discouraging. The total fish landings were almost level; 2.84 million tons in 1979 and 2.92 million tons in 1984, whereas they had increased from 2.2 million tons in 1975. Export value decreased (despite inflation) from US$1.08 to US$0.95 billion. Edible imports increased by 4% in volume, 40% in value; nonedible imports increased by 88% in value. The total value of imported fishery products increased from US$3.8 billion to US$5.9 billion. Total employment in fishing and processing increased from 288,000 to 332,000; changes which probably were associated with decreased productivity per person.

The arguments over the reasons for the U.S. difficulties in fishery development continue but the fishery management program has been described as a process without purpose.(64)

At present, we are doing little more than performing a holding action in most of the fisheries that we purport to regulate. Even that is becoming increasingly difficult as fishing technology improves and more of our resources are exploited to or beyond their limits. Slowing the rate of harvest through imposing inefficiencies on the industry, which has never worked very well, is becoming less effective with every increase in technology.

Yet, use of the underutilized resources remained as a political objective. In 1985 a new 5-year plan for the S-K Grants identified 13 underdeveloped fish resources. Congress increased the industry proportion of S-K funds to 60%, which may be matched by industry up to 50% of the cost of each project.(65)

In fairness to the planners, many of the impediments to development have been identified in the plans. The market survey of 1978 (59) explicitly stated:

The major impediments are organizational, institutional, legal, or motivational in character. Because of these problems a high degree of perceived risks become associated with new development. The degree of perceived risks is compounded by a lack of system level planning capability in the industry. This lack of planning capability is a consequence of the many small regional economic units with little vertical integration that make up a large portion of the industry.

This statement was elaborated at length to explain some of the particulars. A legal factor is the shipping act which requires U.S. construction of fishing and processing vessels, causing not only higher capital costs, but inhibiting experimentation with seabased processing unfamiliar to U.S. businesses. An institutional and motivational factor is the strongly traditional character of the fishing and processing businesses. These are mostly small to medium-sized; few have the capability to plan, finance, and organize new technologies and expanded trade in new products, especially for export.

The federal assistance program appears to perpetuate and encourage the small- to medium-scale activities through optimistic emphasis on technical information, training, and provision of information.

The functioning of the FCMA has provided a preview of future difficulties in both fishery management and development. A review by the recently retired Director of the Office of Fisheries Management of the National Marine Fisheries Service, Mr. Roland Finch,(66) identified the tug-of-war between regional and national interests as a perennial problem. The provisions of FCMA required the regional fishery management councils to recommend decisions to the Secretary of Commerce for review of compliance with national standards and implementation. The views of the councils frequently differed from national policy, as interpreted by the Secretary, and caused a continuing power struggle, which was aggravated by the long time required to resolve the differences.

A major source of the difficulties has arisen from large changes in some stocks, either increases or decreases, due to natural causes rather than fishing. Despite major increases in fishery research, the predictability of such changes remains uncertain. But on the whole, conservation of the stocks has improved. Eight species groups have increased in abundance, 11 remain unchanged, and 6 have decreased.

The development of the fisheries, however, has been essentially limited to increased fishing for over-the-side delivery to foreign processors. The catches of the few vessels able to take part has increased from none recorded in 1978 to nearly 0.7 million tons in 1984. Information on eight major fisheries presented as a "health index" indicates that none have improved by much, and a majority have decreased. Furthermore, the domestic processing industry has improved little, if any.

The problems of overinvestment remain in many fisheries. In some fisheries, less than half the vessels could easily take the allowable catch. The only limitation on entry under the Council system has been applied to the Atlantic surf clam fishery, but so many vessels have been involved--after the limitation--that they can still catch their quotas in only one and one-half days per week. (Limitations on entry to a few fisheries also have been applied by a few states.)

Thus the United States, apparently with nearly 10% of the world's conventional marine fish stocks within its EEZ, has failed so far to really achieve the potential benefits of the fisheries within its EEZ.

NOTES AND REFERENCES

1. Marine Technical Assistance Group, "An Evaluation of Fishery and Aquaculture Programs of the Agency for International Development." U.S. Nat. Acad. Press, Washington, D.C., 161 pp., 1982. See pp. 73-76.
2. A. M. Klausen, "Kerala Fishermen: And the Indo-Norwegian Pilot Project." Universitetsforlaget, Oslo, 201 pp., 1968.
3. G. V. Everett, M. Ansa-Emmin, and I. Mizuishi, A summary overview of fisheries in the CECAF region. *FAO CECAF/TECH/80/21*, 62 pp. (1980)
4. J. I. Clarke, "Sierra Leone in Maps." Hodder and Stoughton, London, 120 pp., 1969.
5. M. K. Hendrix, Technical change and social relations in a West African maritime fishery: A development history. University of Rhode Island, Internat. Cen. for Mar. Resource Develop., *Working Pap.* (21):1-25 (1984).
6. W. F. Royce, H. R. Schmittou, and L. Vidaeus, "Small-scale fisheries development in Sierra Leone." Contract report to U.S. Dept of State from Amer. Tech. Asst. Corp., McLean Virginia, Nov. 1976.
7. FAO, "Yearbooks of Fishery Statistics, Catches and Landings," Annually through Vol. 54 (1950-1983).
8. H. Kitchen, Africa: Year of ironies. *For. Aff.*, *64*(3):562-582 (1985).

9. C. D. May, Sierra Leone: Case of apathy, more than anger. *Int. Herald-Tribune*, pp. 1,2 (June 30 - July 1, 1984).
10. M. A. Robinson and R. Lawson, Some reflections on aid to fisheries in West Africa. *Mar. Pol.*, *10*(2): 101-110 (1986).
11. Statement by Dr. A. C. Bundu, Minister of Agriculture and Natural Resources of Sierra Leone, at the FAO World Conference on Fisheries Management and Development in Rome, 27 June to 6 July 1984.
12. D. Menasveta, Resources and fisheries of the Gulf of Thailand. Training Dept., S. E. Asia Fish. Dev. Cen., Philippines, *Text/Reference Book Series* (8):1-108 (1981).
13. K. Tiews, Bottom fish resources investigation in the Gulf of Thailand and an outlook on further possibilities to develop the marine fisheries in Southeast Asia. *Arch. Fischereiwiss*, *XVI*(1):67-108 (1965).
14. K. Tiews, The German fisheries mission to Thailand [2nd project phase (1962-1964)]. *Veröff. Inst. küsten- binnenfisch. Hamburg*, (36):1-32 (1965).
15. S. Charernphol, Current status of the fisheries in Thailand. *Thai Fish. Gaz.*, *34*(5):525-528, (1981).
16. T. Bhukaswan, Country status report on Thailand. *In* "Symposium on the Development and Management of Small-Scale Fisheries: *Proceedings of the Indo-Pacific Fish. Comm.*, 19th Session, Kyoto, Japan, 21-30 May 1980," FAO, Section III, pp. 247-251, 1980.
17. T. Panayotou, Economic conditions and prospects of small-scale fishermen in Thailand. *Mar. Pol.*, *4*(2):142-146 (1980).
18. Anon. Urgent measures needed to save fishing industry. *Nation Rev.*, Bangkok, pp. 11,12, (Aug. 3, 1981).
19. Anon. Thai fish crisis. *World Fishing*, p. 25. (Jan./Feb. 1981).
20. K. Tiews, Fishery development and management in Thailand. *J. Fish. Res. Board Can.* *30*(12):2333-2334 (1973). Condensation only.
21. R. H. Fiedler, The Peruvian fisheries. *Geogr. Rev.* *XXXIV*(1):96-119, (1944).

22. L. K. Boerma and J. A. Gulland, Stock assessment of the Peruvian anchovy (*Engraulis ringens*) and management of the fishery. *J. Fish. Res. Board Can.*, *30*(12):2226-2235 (1973).
23. T. Wolf, Peruvian-United States relations over maritime fishing: 1946-1969. LOS Inst., Univ. of Rhode Island, *Occasional Pap.*, (4):1-26 (1970).
24. M. Roemer, "Fishing for Growth: Export-led Development in Peru, 1950-1967." Harvard Univ. Press, Cambridge, Massachusetts, 208 pp., 1970.
25. Anon., State takes over Peru fish meal industry. *Fishing News Int.*, (June 1973).
26. T. F. Chaparro, La pesqueria Peruana estudio de caso. *In* "Case Studies and Working Papers Presented at the Expert Consultation on Strategies for Fisheries Development, Rome 10-14 May 1983." *FAO Fish. Report* (295, Suppl.):123-140 (1983).
27. W. P. Appleyard, Peru: A case study in the establishment of a food fish industry for a developing country. *J. Fish. Res. Board Can.*, *30*(12):2236-2241 (1973).
28. Office of International Fisheries, Peruvian fishery policies, 1984. U.S. Natl. Marine Fish. Serv., *IFR-85/21R*, 6 pp. (1985).
29. Statement by Peruvian Delegation led by Emilio Rodrigues Larrain, Vice-Minister of Fisheries, at the FAO World Conference on Fisheries Management and Development in Rome, 27 June to 6 July 1984.
30. M. H. Glantz, Man, state, and fisheries: An inquiry into some societal constraints that affect fisheries management. *In* G. D. Sharp, and J. Csirke, eds., "Proceedings of the Expert Consultation to Examine Changes in Abundance and Species of Neritic Fish Resources, San José, Costa Rica, 18-29 April, 1983." *FAO Fish. Rep.* (291):873-917 (1983).
31. ICNAF (International Commission for the Northwest Atlantic Fisheries), The development of the fishery in the convention area over a series of years. *Stat. Bull.*, *3*:9-17 (1955).
32. 1 UST 477; TIAS 2089.
33. P. Copes, "The Resettlement of Fishing Communities in Newfoundland." Can. Coun. on Rural Dev., 259 pp., April 1972. See pp. 94-95.

34. *Ibid.* p. 77.
35. *Ibid.* Tables 50, 54.
36. *Ibid.* Tables 65, 67.
37. *Ibid.* Preface.
38. ICNAF, *Ann. Stat. Bull.*, 1951-1978. Series continued as *Stat. Bull. of N. W. Atlan. Fish. Organ.*, 1979--.
39. W. G. Doubleday *et al.*, The impact of extended fisheries jurisdiction in the northwest Atlantic. Manuscript report submitted at Workshop to Assess the Implications of Extended Coastal Jurisdiction for the Management and Development of World Fisheries, Seattle, Washington, 8-11 July 1985., 21 pp.
40. E. B. Dunne, Main parameters and characteristics of Canadian Atlantic groundfisheries. *In* "Economic State and Problems of Small-Scale Fisheries." Organ. Econ. Coop. Dev. (OECD), Paris, pp. 5-23, 1976.
41. J. E. Forrest, Development of the east coast fishing industry: Assessment of Canadian government policies. *Mar. Pol.*, *5*(4):294-301 (1981).
42. N. Roy, W. E. Schrank, and E. Tsoa, The Newfoundland groundfishery: Some options for renewal. *Can. Pub. Pol.*, *VIII*(2):222-238 (1982).
43. G. Barrett and A. Davis, Floundering in troubled waters: The political economy of the Atlantic fishery and the Task Force on Atlantic Fisheries. *J. Can. Stud.*, *19*(1):125-137 (1984).
44. *Ibid.* pp. 130-131.
45. C. Lamson and A. J. Hanson, "Atlantic Fisheries and Coastal Communities: Fisheries Decision-Making Case Studies." Dalhousie Univ., Inst. Resource and Environ. Stud., Halifax, Nova Scotia, 252 pp., 1984.
46. U.S. International Trade Commission, Certain fresh Atlantic groundfish from Canada. *USITC Publ.* (1750) (September 1985). Includes reprint of notice of investigation in the *Fed. Reg.* *50*(169):35281-35283 (August 30, 1985).
47. U.S. National Marine Fisheries Service, Weekly Highlights memo, 17-21 March, 1986.
48. S. Moriya, The experience of Japan in identifying and implementing fishery development policies and strategies: Case study. *In* "Case Studies and Working Papers Presented at the Expert Consultation on Strategies for Fisheries Development, Rome 10-14 May 1983." *FAO Fish. Report* (295, Suppl.):92-107 (1983).

49. J. G. Gissberg and Y. Nasaka, Japan's fisheries, 1983. *U.S. Embassy CERP Rep.* (Current Economic Reporting Program) (December 1984).
50. K. Shima, The role of cooperatives in the exploitation and management of coastal resources in Japan. *In* "Case Studies and Working Papers Presented at the Expert Consultation on Strategies for Fisheries Development, Rome 10-14 May 1983." *FAO Fish. Report* (295, Suppl.):243-252 (1983).
51. W. C. Herrington, Operation of the Japanese fishery management system. LOS Institute, Univ. Rhode Island, *Occasional Pap.* (11):1-21 (1971).
52. Education and Information Dept, "The Zengyoren (Summary)." Zengyoren (National Federation of Fisheries Cooperative Associations), Tokyo, Japan, 45 pp., 1985. (In English).
53. Y. Asada, Country status report on Japan. *In* "Symposium on the Development and Management of Small-Scale Fisheries: *Proceedings of the Indo-Pacific Fish. Comm.* 19th Session, Kyoto, Japan, 21-30 May 1980," FAO, Section III, pp. 145-162, 1980.
54. Office of International Fisheries, Japan's salmon supply,1974-1983. U.S. Nat. Mar. Fish. Serv. *IFR-84/79B*, 14 pp. (1984).
55. N. Fujinami, Perspectives of a fishing nation, Japan. Discussion paper presented at EEZ Management Refresher Course, Suva, Fiji, 4-15 February, 1985.
56. National Fishery Statistics Program. Fisheries of the United States, 1984. Natl. Mar. Fish. Serv., U.S. Dept. Comm., *Current Fish. Stat.* (8360):1-121 (1985). (This is the most recent in a series spanning more than 40 years. These statistics differ from those reported by FAO because the catches do not include shell weight of mollusks and are mostly reported as landings which do not include catches sold "over-the-side" to foreign processing vessels.)
57. R. F. Hutton, P. L. Hooper, and R. H. Stroud, United States of America country review. *In* "Allocation of Fishery Resources. Proceedings of the Technical Consultation of Allocation of Fishery Resources held in Vichy, France, 20-23 April 1980." FAO and Amer. Fish. Soc., pp. 613-623, 1982.

58. Bureau of Commercial Fisheries, Trident: A long range report of the Bureau of Commercial Fisheries. U.S. Dept. Int., *Circ.* (14):1-113 (1963).
59. E. R. Combs, Inc., and A. Young & Co., "A Study to Determine the Export and Domestic Markets for Currently Underutilized Fish and Shellfish." Natl. Mar. Fish. Serv., U.S. Dept. Comm., 416 pp., December 1978.
60. E. R. Combs, Inc., "Prospectus for Development of the United States Commercial Fisheries." Earl R. Combs, Inc., Mercer Island, Washington, 441 pp., May 20, 1979.
61. Task Force on Fisheries Development, "Toward a Partnership for the Development of the United States Commercial Fishing Industry," Final Report, U.S. Dept. Comm., 82 pp., May 23, 1979.
62. U.S. Public Law 96-561--December 22, 1980. 94 Stat., 3287-3302.
63. U.S. General Accounting Office, "Uses of Saltonstall/Kennedy Fisheries Development Funds." *GAO/RCED-85-145*. 64 pp., August 30, 1985.
64. J. H. Branson, D. M. Larson, and R. W. Miller, "U.S. Fisheries Managment: Process without Purpose." Speech to Am. Fish. Soc., 1984 Meeting, Ithaca, New York.
65. Office of Industry Services, "Fishing Industry Growth and Development: Long Range Plan for Supporting Private Sector Initiatives through the Saltonstall-Kennedy Grants Program." U.S. Natl. Mar. Fish. Serv., July 12, 1985.
66. R. Finch, Fishery management under the Magnuson Act. *Mar. Pol.*, *9*(3):170-179 (1985).

Chapter 7

The Farmers' Edge

As the world production from wild stocks of fish tops out, the fish supplies for an increasing population must be obtained by increasing the yield per unit area. This is the role of fish farming, a parallel to the role of land farming, which has now almost entirely supplanted the role of hunting for food.

Farmers have an important edge over fishers in even the simplest kind of fish farming. Farmers gain at least a little more control over return on investment in their production facilities and take at least a small step away from the political control of harvesting a resource of the commons. When free of the politics of the commons, fish farm development can be focused on the technical problems of the husbandry and the economic problems of the farming.

Fish farming has been practiced for millenia, but even the present practices range from primitive attempts to protect wild animals at some stage in their lives to total control of an animal's life in an operation comparable to a factory. Such diversity is one of the great advantages of fish farming, because simple techniques can be used by fishers in fresh and estuarine waters to increase the value of their labor in a gradual transition from fishing to farming, and sophisticated techniques can attract major private investments.

Unfortunately, the overall statistics on fish farm production are relatively unreliable. One major reason is inclusion of most transitional fishing-to-farm production in the fish catch data from fresh and estuarine waters. For example, most of the molluskan production is reported as a catch even though a large majority is probably produced on farms. Most freshwater fin fish production is probably augmented by practices that improve the environment for fish, and should be considered as part of fish farming. In addition, major business activity in nonfood fish farming such as bait fish, ornamental fish, and pearl production is usually omitted from national fishery statistics.

The husbandry of fish is extraordinarily diverse, varying as it does with many species, each possibly in many kinds of farming situations. The purpose of this chapter is not to detail the biology or technology that dictates the practices used in the husbandry, but to emphasize the socioeconomic and lifestyle changes that influence public policy. The dominant theme is the transition from fishing dependent on waters of the commons, through the complex arrangements for private use of the commons, to the major types of fish farm businesses. A separate husbandry, technically similar but politically very different, is government enhancement of public fishing waters.

These difficulties mean that few long-term nationwide examples of development can be documented as national case histories as in the last chapter. The examples chosen, therefore, reflect the development problems as they have been identified in plans for an assortment of projects and programs. Furthermore, the chapter has been organized around the predominant social characteristics of the farming systems rather than around the biological characteristics usually used by fishery scientists.

7.1 THE PRIVATE TRANSITIONAL SYSTEMS

The transitional systems are defined here as those that still use the commons during part of the lives of the farm animals or plants, and are thus subject to at least some of the political problems of the commons. These private systems will contrast with the government systems used to enhance the fishing in public waters, which will be discussed in the next section. The private systems, through their use of the commons, are subject to at least some government controls based on other public uses of, or values inherent in, the waters. A major value is, of course, the traditional public right called piscary.

In this as in the following sections on private systems, the term fish farming is used to include all kinds of aquatic animals and plants, whether fin fish or not. The terms farming and husbandry are used rather than aquaculture to emphasize the business and technical aspects, respectively, which range in scale from subsistence level to large-scale transnational activities.

7.1.1 Fishing to Husbandry

Fishers have long identified simple farm practices that would enhance their fish production, and such practices are still widespread in fresh and estuarine waters. Wild fish can be attracted to shelters, possibly through use of food, where they can be caught more easily or as needed. Simple corrals or traps constructed of brush can be used to catch fish during a migration. In very shallow delta lakes, such as those near the mouth of the Nile River, temporary ponds to enclose fish can be constructed by simple dikes. In all such temporary enclosures, the fish can be sorted to remove undesirable species, and they may be fed.

Similar practices are widespread in the floodplain areas of large rivers where fish are found in the ponds and oxbows remaining after floods.(1) Here the fish may be easier to catch than in the open river, and frequently they can be kept in captivity for three to six months, where they can be protected and fed if desirable. Another opportunity for such practices may exist in ponds and small lakes managed for other purposes, such as for farm or village water supplies.

The technology of such fish farming varies with almost every situation depending on the species, on the flow regime in the watershed, on other uses or abuses of the waters, and on the resources or abilities of the farmer.

In most situations such activities are essentially private, even though the waters are common property. They are family or community activities that have been locally accepted as a right or arranged through a lease to an individual (if the laws of the country require it). As with the varied husbandry, the farming practice needs to be varied to fit the local customs.

These activities are transitional, not only in the technical change from capture to culture, but also in the social adjustments of fishers and their families. Fishers frequently have an image of being individualists who live with danger, as with the jangadieros of northeast Brazil who fish in the sea from rafts. They are self-reliant people who have met the challenge of the sea and take pride in doing so. They typically engage in a male-only occupation, and many of them regard farming as an inferior occupation. Fishers may also regard farmers as competitors, and a change to fish farming may cause enmity of former colleagues, especially if the farming depends on collection of wild stocks or fry.[2]

Furthermore, the transition to fish farming may require an investment of time and resources in ponds or enclosures instead of in fishing equipment, which will cause major changes in family activities, responsibilities, and income patterns. And the relationship of a farm operator to employees may be quite different from the relationship of a boat owner to a crew.

7.1.2 Use of Wild Brood Stock

Control over the fish farming operation becomes more certain for many species if reproduction can be controlled. Wild brood stock has been used for millenia and continues to be used when the adults will spawn in captivity, when spawn can be taken manually, or when spawning can be induced by hormonal injections.

Wild stocks are technically less desirable for fish farming, however, because they have evolved to fit the natural habitat, whereas the organisms in captivity are subject to very different conditions. Close confinement enhances the opportunities for spread of disease. In addition, the water regimes in captivity will be different with respect to temperature, currents, and chemistry, all of which may affect the behavior and consequently the growth of the fish. Domesticated stocks, on the other hand, may be selected for disease resistance and adaptation to the conditions of husbandry.

In many species the adults cannot be induced to spawn in captivity because they require bottom conditions, current conditions, or other stimuli that cannot be reproduced in the controlled waters. In some cases the spawning may be successful, but the eggs and fry then cannot be nurtured satisfactorily. In either situation, an alternative to the use of adults may be the capture of the young as described in the next section.

Although farming of many choice wild species has been attempted, few operations have been economically viable. One example is the difficulties that have been encountered in attempts to culture several species of flatfishes which, as a group, constitute one of the major fish resources of north temperate waters. Government efforts in both Europe and North America started in the early twentieth century to hatch eggs and obtain fry for stocking. The fry can be easily produced in immense numbers but the practices have never been shown to be economically effective. More recently, extensive efforts have been made to perfect techniques of farming, especially through use of heated effluents which should increase the rate of growth. But so far, commercial farming does not appear to be practical because of the technical complexities as well as the high cost of feeding these carnivorous fish.[3]

In addition to technical difficulties which will vary with every species, external social difficulties must also be expected. Farmers need to obtain brood stock (a public property) from the commons in competition with fishers, which may be uncertain if the fishery is intense. The farmers may compete with the fishers for markets and thus earn their enmity, especially if the farmers are perceived to be more affluent.

7.1.3 The Commons as a Nursery

A large fraction of the world's fish farming depends on "seed" caught in the commons.[4] The organisms sought include finfish, mollusks, crustaceans, and plants; the collection places are primarily estuaries and rivers. The system is feasible with many organisms which produce very large numbers of spores or fry, the removal of some of which will scarcely affect the population remaining. The practice allows species to be farmed when the spawning of adults in captivity cannot be obtained or the culture of the fry cannot be managed. The latter case is especially difficult with species that have tiny fry which must have natural foods and special rearing techniques to survive. Techniques of inducing spawning with injections of hormones and of feeding tiny fry with live plankton have been discovered for many species, but the traditional methods for these may still be used by less sophisticated growers. Several dozen species are farmed with wild seed, but only the major ones will be discussed here.

A major example is most of the oyster farming in the world. This takes place in estuaries where almost all the farms operate under a government license or lease. The oyster larvae are produced in huge numbers by natural spawning of both farmed and wild oysters; they drift for a time with the currents and become attached as spat to hard objects, usually old shells. Then the husbandry begins with collection and placement of the shells with the spat attached. The oysters depend completely on natural food, which they filter from the public waters.

Oyster farming remains in competition with oyster fishing in many localities; the eastern United States is a good example which has been summarized recently.[5] The oyster harvests from southern New England south to the Carolinas, most of which were from wild stocks, were large enough to supply some 35% by value of the total fish landings in the United States in the latter part of the nineteenth century. The fishing was also an important source of employment for oyster fishers, the "baymen." Private planting on leased areas began after 1860 and spread, despite continued opposition by baymen who forced State limitations on the size of leased areas.

The Chesapeake Bay oyster fishery, divided between the states of Maryland and Virginia, is an example of the makeshift political compromises between use of oyster grounds for public harvest and for private oyster farming. The fishing on public grounds has long been restrained in order to divide the catch. The individual fishers recently have been allowed to use only tongs, a heavy pincerlike manually-operated device with which they grope for oysters in depths up to about 6 m. The oyster farmers on leased grounds can, of course, use dredges towed by powered vessels to harvest the oysters as efficiently as practical. Interestingly, dredges were also permitted recently on grounds open to public fishing in Maryland provided they were towed by the traditional sailing vessels, although this was modified at one time to permit powered vessels only on Mondays and Tuesdays!

These restraints have been political compromises justified by the need for conservation and have been accompanied by large annual public expenditures for oyster planting and enforcement of regulations. The long-term result has been a failure of the fishery management program to sustain the yields; production of oyster meats from Chesapeake Bay, which was around 50,000 tons annually in the 1880s, recently has been about 10,000 tons.

Private oyster production, however, is becoming less dependent on natural seed since discovery of reliable laboratory methods of spawning oysters and rearing larvae in the 1960s. The seed supply need not be a problem now, but the political problems of dividing the public grounds between fishers and farmers remain.

Governments will continue to have a major impact on oyster farming as well as fishing through measures taken to protect public health, which will include designation of growing areas, establishment of standards for healthful products, and protection of the growing areas from pollution. The latter appears to be a limiting factor for expansion of oyster farming in many parts of the world as the coastal waters are increasingly polluted. In fact the worldwide production of oysters appears to be topping out as it has increased only from about 0.8 million tons annually in the 1960s to about 1 million tons in the early 1980s (FAO statistics).[6]

Another major fish farm business dependent on the commons as a nursery is that for milkfish in several countries of Southeast Asia. The milkfish is a quality foodfish that lives along the coasts and in rivers where it feeds almost entirely on algae. It is a large fish, weighing up to about 20 kg, that broadcasts eggs in huge numbers in warm coastal and estuarine waters where they hatch in a few hours and the fry can be easily collected. After capture, they are placed in ponds suited to the advanced fry stage, the fingerling stage, and later the production stage. Ponds are usually constructed near tidal streams or in coastal swamps where they are managed by control of water circulation and by fertilization to produce the algae eaten by the milkfish. The larger fish may also be fed artificial diets until they reach market size of about 400-500 g.

Unfortunately, the adult milkfish will not spawn naturally in captivity, nor has it been possible yet to induce spawning consistently by hormonal injections. Otherwise the species is nearly ideal for farming because, in addition to being readily marketable, it is disease resistant and cheap to feed. The limiting factor in production, recently about 200,000 tons annually, is already the availability of fry in many localities. It will remain the limiting factor unless a breakthrough occurs in control of the spawning.(7)

A group of fish with habits similar to milkfish are the mullets which are grown extensively in southeast Asia and in the Mediterranean region by husbandry similar to that used for milkfish. These fish have widespread promise for farming because one species or another occurs in almost all coastal tropical and subtropical waters (occasionally in temperate waters), and for some species the reproductive process has come under control.(8)

The major fish farming business in the world, however, has been based on the Asian carps, which until recently depended entirely on fry caught in large rivers. This group includes the common carp that readily spawns in ponds, and which has been domesticated for centuries throughout much of Asia and central Europe. But several other choice food species, including the silver carp, bighead, grass carp, black carp, catla, rohu, and mrigal, spawn in rivers and will not spawn in ponds. The fry for these species has been collected for centuries in the rivers by farmers or seed suppliers. Recently, however, techniques for inducing spawning have been perfected and are widely used in China, India, and other countries of Southeast Asia (see Section 7.3.2).

Much of the shrimp farming business of the world depends on trapping young wild shrimp. Most of the marine shrimps sought for food spawn in the sea near estuaries to which the larvae migrate and live for several weeks or months. Some of the simplest shrimp farming in southeast Asia is based on ponds excavated in coastal swamps which are constructed to allow tidal waters to enter with young shrimps which are later trapped for the market as they attempt to return to the sea. Many of the operations also profit from fin fish which enter and grow in the ponds. A few, where completely fresh water is available, may grow rice and freshwater shrimp together in the paddies or as alternate crops.[9]

Alternatives to passive trapping are to capture the postlarval shrimp with dip nets, traps, or small tow nets. A variety of methods have been developed for various localities and species but may be expensive, uncertain, or cause injury to many young shrimp.

This business as a whole is changing as techniques for controlling the spawning and larval development are perfected and put to use. The Japanese discovered the methods for one of their choice species in the 1930s and had applied them widely by the 1960s. Further improvements have been made recently by the Taiwanese. Now the farming is so attractive that efforts are being made in many countries to bring the husbandry of other species, especially the large freshwater shrimps, under complete control and reduce reliance on the commons.

The growth of the shrimp farming business in Ecuador provides an example of the limitations of the commons. The country has a large natural shrimp producing area in the Gulf of Guayaquil around which shrimp farms sprouted in the late 1970s. Shrimp harvests increased from wild catches of about 6000 tons in 1975 to about 37,000 tons in 1983 due to farm production from about 45,000 ha of ponds. The farm operations have depended on wild postlarvae caught by artisanal fishers whose numbers had grown to an estimated 90,000 in 1984. The business has attracted substantial amounts of private capital, and the country has benefited from export of a large majority of production. By 1985, however, the production appeared to have peaked because of a shortage of postlarvae. Hatcheries were being constructed by the government and private firms, but the business of consistent high-quality hatchery production of postlarvae has not been perfected and some of the farmers fear a crisis due to a shortage.[10,11]

Seaweed farming in eastern Asia, where more than 1 million tons are produced annually, is another major activity that has been dependent on the commons for a nursery--in this case spores. The seaweeds are species of red, brown, or green algae which reproduce in a two-stage cycle. The leafy plants or sporophyte phase grow during the winter and produce sexual products in the spring that drift with the waters. After sexual fusion, carpospores are produced which become attached to shells where they mature during the summer and produce monospores in early autumn that also drift with the currents. These become attached to shells or to special nets or ropes used in the intensive husbandry. The life history of the plants has been studied intensively and laboratory production of both carpospores and monospores is being rapidly adopted.[12]

The use of the commons as a nursery continues with the above organisms as well as with many other farm species. Included are mussels, clams, marine catfishes, eels, smelts, and Japanese yellowtail. It continues when it is not possible to control spawning artificially or when it is more economical to collect wild seed that it is to grow it on farms. But complete control of the reproductive process has such great advantages in selective breeding as well as in reducing problems of using the commons that major efforts are being made to completely domesticate more and more species.

7.1.4 The Commons as a Ranch

The anadromous species of fish, especially the salmonids, sturgeons, and shads that return consistently to a home stream, offer the possibility of using the commons to avoid expensive confinement and feeding during farm operations. The practice has been called ranching because of its similarity to the land farm operations of supporting livestock by grazing on uncultivated pasture land. Salmonid ranching is a widespread public practice (see Section 7.2.3) around the northern temperate latitudes and is growing in southern South America, New Zealand, and Australia.

Private salmon ranching, however, conflicts with the traditional perception of rights in ocean fishing. Its recent history in the United States revealed the basic political problems. State legislation to permit private salmon ranching was enacted in California (1968) and Oregon (1971). In addition, Alaska enacted legislation in 1974 to permit private but nonprofit salmon ranching, which was to permit operation of fisher's cooperatives.

A proposal in 1975 to allow private salmon ranching in the state of Washington was defeated because of the vigorous objections from the existing salmon fisheries based on expectations that such ranching would favor large corporations and would eventually modify the concept of salmon as public property. Private ranching in Oregon and California did start in the early 1970s with large corporate investments and technically competent organizations. But constant political pressure and interference from fisher's organizations has contributed to failure of most of the businesses to reach a profitable stage by 1986.[13]

7.2 PUBLIC ENHANCEMENT OF THE FISHERIES AND IMPROVEMENT OF THE COMMONS

Although the public aquacultural programs are not fish farming in the sense of a business, they operate in one form or another in many countries, and they either set examples for private farm operations through demonstration of the husbandry, or have developed a scientific basis for farming through research on basic biological and engineering problems. They are, of course, primarily to support fishing in the commons through stocking of desirable species. Improvement of the aquatic environment is also a government activity which, along with aquaculture, has enhanced production from the commons.

This section is a brief summary of public aquacultural policies and their relationship to private fish farming.

7.2.1 The Aquatic Environment

Every major river system in the world is being modified by agriculture in its drainage basin, by deforestation, by control of the water through diversions or dams or levees, and by pollution. Any of these factors will cause changes in the natural fish populations that have supported fisheries. Governments, therefore, do many things to minimize ill effects and, in most cases, try to improve the environment for fishing.

Fisheries are now considered as a major part of the environmental concerns; indeed the presence of fish is an index of water quality. The measures taken to protect them and enhance their environment are analogous to the measures taken to foster livestock on land farms and can be considered aquacultural. Most inland and coastal fishery management is now extensively involved with environmental quality.(14)

The policies for such environmental protection became widespread during the 1970s as procedures for environmental impact assessment were adopted by many nations and as environmental criteria were adopted for siting of industry. The major multilateral development banks and the United Nations programs adopted in 1980 a "Declaration of Environmental Policies and Procedures Relating to Economic Development."(15) This reaffirmed the principles adopted at the UN Conference on the Human Environment in Stockholm in 1972 and pledged the development organizations to integrate appropriate environmental measures in the design and implementation of economic development activities.

7.2.2 Stocking

Most governments stock fish as an attempt to compensate for environmental change or as an attempt to enhance fish stocks in natural public waters. Recent experiences in Europe and the United States are illustrative.

Thirteen member nations of the European Inland Fisheries Advisory Commission (EIFAC) reported at a 1982 meeting that stocking was a common practice.[16] All types of water bodies and at least 30 species were involved. Part of the stocking was a transfer of live fish from one water body to another, but most was from hatcheries. Most of the stocking enhanced fishing for recreational fishers whose numbers have been rapidly increasing (see Section 8.1).

The United States has supported large public stocking programs since about 1880. As in Europe, some of the stocking is transfer of wild fish, but an extensive hatchery system is supported. The federal government operated 89 hatcheries in 1982 which stocked a total of some 311 million fish of 49 species weighing about 2800 tons.[17] The states also operated about 400 hatcheries, so it may be estimated that the total government aquacultural production is currently about 10,000 tons. In addition, some privately operated hatcheries are also stocking public waters as part of mitigation agreements with governments over environmental damage to the fisheries.

Stocking policies in both Europe and the United States are popularly supported, especially by the large numbers of recreational fishers. Numerous scientific studies have been made of particular stocking experiments, but scientists in both Europe and the United States have reservations about the long-term efficiency of overall programs, especially those intended to support commercial fisheries.

7.2.3 Ranching

All of the governmental stocking programs are ranching, in the sense of releasing animals to graze in the commons and eventually be caught, but the term *ranching* has been used especially for the release of salmon to "graze" in the ocean. This salmon culture has resulted in spectacular increases in salmon runs, but the economic efficiency has not been reported.

After the widespread trend toward extended jurisdiction over coastal fisheries and the special consideration being given to anadromous fish, it became clear that high-seas fishing for salmon would be restricted. The Japanese government, therefore, moved to expand salmon ranching in the 1960s and 1970s. The government built hatcheries near many of the salmon streams in Hokkaido and northern Honshu, and also financially supported some hatcheries operated by private institutions, including cooperatives.

The catches of chum salmon, the principal species in the coastal waters of Japan, had risen to an annual average of about 20,000 tons in the 1880s, which apparently was more than the sustainable yield, because the catch declined and averaged only about 10,000 tons from about 1895 through the 1960s. Returns from the hatchery program grew rapidly in the 1970s and have since increased the coastal chum production to 123,000 tons in 1983. The cost to the government in the early 1980s was equivalent to about US$15 million annually and was in addition to costs borne by coastal fishers.[18]

The Japanese had studied carefully the techniques used in the United States where salmon hatcheries had been operating since late in the nineteenth century. Public hatcheries had been built in the states of California, Oregon, and Washington, and private hatcheries had been required to be operated by Alaskan salmon companies. Other hatcheries had been started in British Columbia, Canada, but doubts about the efficiency of salmon hatcheries arose and the Alaskan and British Columbian hatcheries were abandoned in the 1930s.

Careful examination of the hatchery techniques in the 1950s and 1960s revealed serious flaws in the methods which, after correction, led to reorganization and expansion of the salmon hatcheries in California, Oregon, and Washington, and a new program in British Columbia. The programs have been supported largely as mitigation for loss of natural spawning and nursery areas for coho and chinook salmon, species favored by both commercial and recreational fishers.

The improved techniques also were adopted in a salmon ranching program of Alaska, which started in the early 1970s and which has been credited with steadily increasing the commercial production of pink and chum salmon above natural returns by 3600 t in 1980 and by 16,200 tons in 1984. This program is operated by nonprofit corporations under special license from the State for the benefit of all commercial fishers in the enhanced fisheries.[19]

7.3 THE PRIVATE FORAGE SYSTEMS

The largest and oldest fish farming systems in the world are those integrated with land farming in ways that are mutually beneficial. They have developed as a farmer's art, especially in China, India, and the countries of Southeast Asia. They are based primarily on species of fish that have come under complete private control, on polyculture of more than one species at a time, and on feeding systems that recycle wastes from land farming operations.

The favorite species eat plants or plankton or benthos and prefer warm, fresh pond waters. They tolerate relatively low dissolved oxygen levels and relatively high alkalinity levels, so many different freshwater environments are suitable and ponds can be fertilized to produce food for the fish. The major species are the carps and related species of the Cyprinidae, the several species of tilapia of the Cichlidae, and other species that can frequently be included with benefit to the system.

The fish farming systems have become remarkably diverse as they have been developed over many years to fit into the lives of the farmers. They also have received intensive scientific research in recent decades directed at increasing the potential production per unit area.

7.3.1 Established Systems

Chinese fish farmers have been developing their art for millenia; a treatise on farming of the common carp was published in 473 B.C., and writings of the following centuries refer to the widespread practice until the seventh century A.D., when political problems arose. The Emperor, with the same name as the common carp (Li), banned consumption of his namesake, whereupon the present polyculture systems using the Chinese "family" carps began.[20]

The four family carps, still the major species used in China, are riverine fish that spawn naturally only in running water. They are the grass carp, black carp, silver carp, and bighead carp. All of them, as well as about ten other species, share a high tolerance for the water conditions of intensively fertilized ponds and get along well together. The silver carp feeds especially on phytoplankton; the grass carp on large plants; the bighead carp on zooplankton; the black carp on snails; the common carp, goldfish, and mud carp on bottom detritus. They also tend to live in the water layer where they feed.

The fish farming systems vary according to the waters; extensive systems in lakes and reservoirs may add 50-100 kg/ha/year to the natural production, whereas intensive pond culture produces an average of 2750 kg/ha/year. Pens and cages are used in lakes and rivers, frequently with supplemental feeding, and in some waters produce an average of 16 kg/m^2/year.

The variety of species and their tolerance of varied water conditions permit many kinds of farmers to include fish as a profitable crop in almost any kind of a farm pond. Some of the most intensive pond systems are integrated with pig and poultry production in which the manures are used to fertilize the ponds. Silkworm rearing and various field crops also produce wastes which either feed fish directly or fertilize the ponds. Kitchen and human wastes are also used. In return, the pond bottoms accumulate organic silt which can be used to enrich the soil for field crops.

The ponds are also integrated into the farm layouts. The ponds may store water to be used for irrigation or livestock watering. The pond dikes can be planted with crops or desirable trees that will protect the dikes from wave erosion during storms.

All of these practices represent development of the art which is now being enhanced by government assistance in research and education. A major breakthrough on the problem of supplying seed occurred in the 1960s when control of the reproduction of the river spawning carps was achieved by hormone injection. The seed are now produced almost entirely by seed farms.

The freshwater fish farming in China utilized some 2.74 million ha of the water area in the country for a production of about 0.8 million tons in 1979. This was an increase from a level of about 0.3 million tons annually in the 1950s. Such farm production is, however, only a part of the fish supply which recently has totalled about 4-5 million tons. China also has an extensive coastal marine fishery, marine fish farming, and freshwater fishing which, although it recently has ranked third among countries in total fish production, bring its fish supply to only about 5/kg/person/year for its population of about 1000 million people.

7.3.2 Transfer of an Art

The Chinese art of fish husbandry is providing a model for other countries with large freshwater areas, but even where the same species can be used, the main restraint on development appears to be adoption of the art by the farmers. An example is a recent effort in Bangladesh, a country embracing delta areas of three large rivers, but which probably cannot expand significantly either its freshwater or marine fishing. Yet it has an estimated 87,000 ha of small ponds which have been producing an average annual yield of only 115 kg/ha. Such ponds managed with proven polyculture technology should easily produce an average of at least 500 kg/ha/year.

The constraints to such farm development have been identified as including water management problems such as drying or flooding, weed control, prevailing patterns of multiple water use, complex and absentee ownership, lack of experience among pond operators, lack of effective extension services, and short-period leasing by governments.(21)

This is a situation in which any technical change, if it is to increase fish production, will create a need for local social changes. Promotion of the development by government will require special institutional sensitivity to local values,(22) even though the integrated farming will probably greatly increase the farm productivity after it has become established.(23)

Similar difficulties have apparently been encountered in many projects sponsored by multilateral and bilateral donors during the 1970s during which time a total investment of about US$685 million was identified.(24) Although part of the projects supported intensive fish farming development, most of the investment was in support of small-scale fisheries in poorer countries. Individuals from 22 federal, international, private, or academic organizations involved with the investments, when asked about the problems associated with execution of the projects, cited managerial or economic problems in 22 instances, social or legal problems in 16, and biological or engineering problems in only 7.

7.4 THE PRIVATE FEED SYSTEMS

Although supplemental feeding is practiced in many of the forage systems, its use is usually an art based on experience with the pond management, the combination of species, and the role of the fish among the other farm crops. In contrast, the feed systems have usually become specialized in the production of a single species that is fed a complete formulated diet and is crowded in the water as much as is practicable. Such systems are more likely to be managed by people who manage only a fish producing system; who are accustomed to handling money and credit; who understand the biology, nutrition, and diseases of the fish; who can prepare a product that will meet national food standards; and who will seek to evaluate the usefulness to them of improved farm technology.

The feed systems have made it possible to farm carnivorous fish which need high protein diets and which include many of the choicest food and sport species. Were these fish fed their natural diets, they would need 7 kg or more of food (wet weight) to add 1 kg of weight. Since their natural food is other aquatic animals, a supply of such quantities is practical only if suitable scrap fish or other low-cost animal flesh is available or if a nutritionally complete formulated diet can be manufactured. Such diets have been discovered through nutritional research and have become available from feed manufacturers. Their use is usually more efficient, because less than 2 kg (dry weight) may be needed to add 1 kg of (wet) body weight.

The feed systems also require more careful management of water quality. Whereas the productivity of the forage systems is limited by the recycling of organic materials--while maintaining suitable water quality for the fish with little if any flow, the water supply for feed systems must be arranged to supply suitable temperature, dissolved oxygen, and minimal accumulation of fish wastes. This may be done by confining the fish in pens or cages through which fresh or estuarine waters can circulate or in tanks or small ponds through which water flows.

Some feed systems, therefore, still are dependent on waters of the commons. The pens and cages are almost always in public waters where their operation is subject to government permission. The tanks or ponds are usually located near a stream or an underground source where long-term water rights of a farmer are usually clearly defined and publicly accepted.

A modified flowing water system occurs in some pond fish farming where water quality can be closely monitored and corrected as necessary with occasional water changes or aeration. This is practiced with ornamental fish and some pond fish such as catfish.

The farming of many species of food fish has been attempted in feed systems using wild brood stock or wild seed. A few have been successful, notably yellowtail in Japan, but the most promising are the systems which have become independent of seed acquired from the commons.

7.4.1 Catfish Farming in the United States

Catfish production, which reached 109,000 tons in 1984, has been the fastest growing farm business during the past three decades in parts of the Mississippi Valley area of south-central United States.(25) Its history is an illustration of the integration of technical solutions into farm management practices.

Most of the production is of channel catfish, a species native to the area. It was known as a choice food fish during the 1950s when it and all other members of the catfish family provided a catch from lakes and large rivers of only about 5000 tons annually. Most farm production still comes from the lower Mississippi Valley area, although some is produced elsewhere in the warmer parts of the United States.

Commercial fish farming in the area started from the experience gained in managing warm-water ponds for subsistence and recreational fishing.[26] Such management was based on natural forage systems and mixed populations of species in impoundments or ponds used for water power, irrigation, or livestock. The earliest commercial production from these systems was, however, monoculture of golden shiners or fathead minnows sold to sport fishers for bait or goldfish sold in the aquarium fish trade.

Fish farming became commercially attractive because of crop acreage controls on rice and cotton, large areas of low-lying lands that could be flooded, reservoirs being constructed for irrigation, duck shooting areas in flooded timber lands, and the need for a crop rotational system in rice farming.[27] The rotation of fish and land crops was practical because the flat land could be easily diked, the soil would retain pond water, and good quality water could be pumped from the water table only a few meters below the surface. Such a water supply was free of fish disease organisms as well as seed of undesirable species.

The management of the system steadily evolved as profits from different crops could be anticipated; as methods of controlling catfish reproduction were discovered; as water quality, nutrition, and disease research provided improved techniques; and as pond engineering provided better methods of pond construction and water management. Most of the production is now from a true feed system based entirely on catfish fed a formulated diet after the early juvenile stages.

The farm systems were not complete, however, without attention to the product and markets. Flesh quality improved through knowledge of its relationship to diet and water quality; market research provided knowledge of preferred market sizes and products; and specialized processing businesses began to operate.

The catfish farm businesses had had periodic problems matching production to demand as production increased about tenfold after the early 1970s. Now the demand for catfish may jump even more as the product has become nationally known and is about to be offered on the menus of restaurant chains. Church's Fried Chicken Inc. plans to offer it in each of its 1,514 restaurants, and the Long John Silver seafood restaurant chain plans to increase its use.(28)

Many of the farmers still have alternatives of raising catfish, rice, cotton, or soybeans on their land according to market conditions. And they may also be able to return to the earlier business of baitfish, the production of which grew from an estimated 10,000 tons in 1980 to 21,000 tons in 1984. Such flexibility appears likely to result in continuing economic development of this fishery sector.

7.4.2 Salmonid Farming

The salmons and trouts are choice fish to eat and especially amenable to farm husbandry if their requirement for clean, cool, well-oxygenated water is met. They require rapidly changing water to maintain oxygen levels and carry away wastes, with temperatures preferably between 10 and 15°C. All of them produce relatively large eggs and larvae that will eat artificial food. They easily yield eggs and sperm to manual methods when they are "ripe" and the fertilized eggs can be easily hatched in incubators.

The public interest in salmon and trouts has led to extensive research on their physiology, nutrition, and diseases. A few species, notably the rainbow trout, have been selectively bred to produce varieties that resist diseases, grow rapidly, and spawn at different seasons, which can then supply a continuous production cycle.

The private trout farming in Europe, North America and Japan originated from the public programs that began during the nineteenth century. In the United States, private trout farming began in the 1850s soon after news of the techniques for producing seed arrived from Europe. The first "hatcheries" tried to raise trout for the market, but most of the profits came from sale of fertilized eggs or brood stock to government rearing stations which were expanding rapidly. A century later, the private trout farmers were producing only about 1000 tons annually.[29] Then nutritional research led to production of improved diets which became cheap, in part because of the rapid increase in supplies of fish meal (see Section 6.4), which provided proper levels of the critical amino acids. Food pellets, which did not fall apart in the water before consumption, became widely available to trout farmers and production jumped to about 9000 tons by 1968,[29] and then to about 23,000 tons by 1984.[25]

The trout farms in the United States are limited by the supply of suitable water. Most of them are small operations scattered throughout the country where suitable cool spring water is available. Large-scale operations are, however, possible in southern Idaho, where some extraordinary springs produce large volumes of water at a temperature near 15°C the year around, an almost ideal temperature for rainbow trout rearing. Here the waters can be loaded with up to 10% fish by weight, and raceways can produce up to 500 tons/ha/year. With such volumes, a large farm can have a railroad spur to supply feed grains to its own mixing plant; it can produce quality products with fish killed and frozen within minutes and packaged for home or restaurant trade.

Trout farming has also expanded rapidly during recent decades in Japan and several European countries. Where spring water supplies are not available, a technique of rearing rainbows in saltwater pens has been developed. And where scrap fish or low-cost fish such as capelin are available, diets can be cheap. European production in 1981 was estimated at about 90,000 tons, with that of the Japanese at 20,000 tons.[30]

Private salmon farming, with more recent business growth than trout, has exploded during the past decade. Small operations have existed for a century or more to supply seed for governments or private streams in a few countries, and the salmon market has been supplied principally by either fresh or canned products from fishing. But wild supplies of both Atlantic and Pacific salmons have topped out, and demand for frozen salmon has burgeoned.

The farming technique, which is based on cage culture in estuaries, is practical where the seawater temperature does not rise above about 20°C or fall below 5°C for very long in winter, and (as with trout) where low-cost diets are available. The principal species are the Atlantic salmon and the coho salmon of the Pacific. Most of the practices concerned with handling of the fish, preparation of diets, control of disease, and preparation of products for market are adaptations of the methods used with trout.

A major restraint is the need to use the commons where either commercial fishers or shoreside residents may object, or where waters may be polluted. But the public has received salmon cage farming more favorably than private salmon ranching, especially in Norway.(31) Norway has had acute social problems among the fishers in isolated communities caused by the failure of fishery management in the northeast Atlantic. Now the country considers Atlantic salmon (and trout) farming as an alternative to small-scale fishing. It is encouraging small-scale salmon farms, each limited in most places to a maximum cage capacity of 8000 m^3, with financial and advisory assistance from the government.

Production of farmed salmon in Norway began in the late 1960s, reached 2000 tons in 1976, 10,000 tons in 1982 (when 215 farms raised salmon), and is projected to approach 80,000 tons in 1990. By producing a high quality, uniform product available the year around, the farmers are creating a new market as well as new income for a distressed area and an important export product for the country. Even restaurants in Seattle, Washington, which has long considered itself to be the salmon capital of the world, have begun to feature Atlantic salmon from Norway.

Other countries are far behind Norway in salmon farming but will probably increase their production rapidly. Japan produced about 2900 tons, United Kingdom 2500 tons and the United States about 900 tons in 1983. Other northern European countries, Canada, Chile, and New Zealand have small but growing operations, and Australia is expected to start soon with operations in Victoria and Tasmania.(32)

7.4.3 Ornamental Fish

No comprehensive statistics are available on the world's ornamental fish business, but glimpses available from scattered sources suggest that it is very large in value if not in weight. For instance, in the years 1970-1972, the FAO statistics on international trade in conventional fishery products show an import value of US$3.3 billion to US$4.5 billion. At about the same time an estimate of the retail value of the live ornamental fish trade including accessories was at the same level--US$4 billion.

A recent review(33) points out that a large proportion of this trade is in wild fish, but that catches from many localities appear to be near the limit for many species. The alternative is farm production, which appears to be a substantial business in several countries. Estimates for the United States published in 1971 indicated there were about 450 farms and import firms which handled more than 200 million fish sold through about 6000 pet shops and 10,000 variety stores.(34)

More and more choice species that had been available only from wild stocks have been bred in captivity and are being supplied by farmers. Hong Kong, Thailand, and other southeast Asian countries are exporting large quantities of farmed aquarium fish, some of which come from species native to Latin America.

The farming technology is complex, because each of the hundreds of species, from either fresh or salt water, have individual food and habitat requirements that must be met if they are to be bred and reared in captivity. Yet some choice species such as the oscar will breed almost continuously, and one pair might produce young valued at US$3000 per year (at 1969 prices).

The prospects for further expansion of the business appear to be good, but the risks for small-scale farmers also appear to be great. Many countries have been controlling imports for fear of introducing fish diseases or exotic species into local waters. Inspection systems for both exports and imports have been started. Licensing and permit systems for producers and dealers are now in place in many countries and are expected to expand. Production for international sale will require careful quality control and good market information.

NOTES AND REFERENCES

1. R. L. Welcomme, Fishery management in large rivers. *FAO Fish. Tech. Pap.*, (194):1-60 (1979).
2. R. B. Pollnac, S. Peterson, and L. J. Smith, Elements in evaluating success and failure in aquacultural projects. *In* L. J. Smith and S. Peterson, eds. "Aquaculture Development in Less Developed Countries: Social, Economic, and Political Problems." Westview Press, Boulder, Colorado, 152 pp., 1982. See Chapter 12, pp. 131-143.
3. J. F. Bardach, J. H. Ryther, and W. O. McLarney, "Aquaculture: The Farming and Husbandry of Fresh-water and Marine Organisms." Wiley-Interscience, New York, 868 pp., 1972. See Chapter 31, pp.575-586.
4. The term "seed" is widely use in fish farming to mean fish fry at a stage when they can be identified and when they can be readily transported in large numbers.
5. J. L. McHugh. Fishery Management. *In "Lecture Notes on Coastal and Estuarine Studies," 10*:1-207, Springer-Verlag, Berlin. See Chapter IV, pp. 64-67
6. J. F. Bardach *et al.*, *Op. cit.* Chapter 36.
7. *Ibid.*, Chapter 17, pp. 313-349.
8. *Ibid.*, Chapter 16, pp. 285-312.
9. *Ibid.*, Chapter 32, pp. 587-632.
10. Office of International Fisheries, Ecuadorean shrimp culture and exports. U.S. Natl. Mar. Fish. Serv., *F/M321:DW, IFR-85/50R1* (1985).
11. Office of International Fisheries, Ecuadorean shrimp hatcheries. U.S. Natl. Mar. Fish. Serv., *F/M321:DW, IFR-85/3-B* (1985).
12. J. F. Bardach et al., *Op. cit.*, Chapter 42, pp. 790-814.
13. T. Y. Nosho, E. O. Salo, and J. Dee, eds., Ocean ranching in Washington: A workshop summary. *Wash. Sea Grant Rep. 75*(1):1-24 (1975).
14. T. Petr. ed., "Summary Report and Selected Papers Presented at the IPFC Workshop on Inland Fisheries for Planners," Manila, The Philippines, 2-6 August 1982. *FAO Fish. Rep.*, (288):1-191 (1983).
15. UNEP, Environmental impact assessement: A tool for sound development. United Nations Environmental Programme, *Ind. Environ*, Spec. Iss. (1):1-20 (1980).

16. European Inland Fisheries Advisory Commission, Report of the symposium on stock enhancement in the management of freshwater fisheries. *EIFAC Tech. Pap.*, FAO, (42):1-43 (1982)
17. U.S. Fish and Wildlife Service, Propagation and distribution of fishes from national fish hatcheries for fiscal year 1982. *Fish Distrib. Rep.* (17) (1982).
18. U.S. Office of International Fisheries, Japan's salmon supply, 1974-1983. U.S. Dept. Commerce, Natl. Mar. Fish. Serv., *F/M321:GH, IFR-84/79B* (1984).
19. Personal communication from J. Hansen of the Alaska Dept. of Fish and Game (1985).
20. FAO, Freshwater aquaculture development in China, *In* "Report of the FAO/UNDP study tour organized for French-speaking African Countries, 22 April - 20 May 1980." *FAO Fish. Tech. Pap.*, (215):1-124 (1983).
21. South Asia Projects Department, Bangladesh. Internal Report of The World Bank, Fishery Subsector Review, December 22, 1983.
22. M. S. Khan, Culture fisheries of Bangladesh: The issue of unused ponds. *In* T. Panayotou, ed., "Small-Scale Fisheries in Asia: Socioeconomic Analysis and Policy." Internat. Dev. Res. Cen., Ottawa, Ontario, pp. 261-268, 1985.
23. Y. C. Shang and B. A. Costa-Pierce, Integrated aquaculture-agriculture farming systems: Some economic aspects. *J. World Mariculture Soc.*, *14*:523-530 (1983).
24. L. E. Grivetti, The rationale, promise, and realities of aquaculture: A cultural-nutritional perspective. *In* L. J. Smith and S. Peterson, eds., "Aquaculture Development in Less Developed Countries: Social, Economic, and Political Problems." Westview Press, Boulder Colorado, 152 pp., 1982. See Chapter 2, pp.11-20.
25. U.S. Department of Agriculture. Preliminary data.
26. H. S. Swingle, History of warmwater pond culture in the United States. *In* N. G. Benson, ed., "A Century of Fisheries in North America." Amer. Fish. Soc., Bethesda, Maryland, Sp. Publ. (7):95-105, 1970.

27. A. H. Hulsey, Trends in commercial fish farming practices in Arkansas. Paper presented at the eighteenth annual conference of the Southeastern Association of Game and Fish Commissioners, Clearwater, Florida, 18-21 October, 1964.
28. J. Valentine, Catfish demand explodes since fillets reached Church's fried chicken menu. *Wall St. J.* (8 July 1985).
29. J. T. Bowen, A history of fish culture as related to the development of fishery programs. *In* N. G. Benson, ed., "A Century of Fisheries in North America." Amer. Fish. Soc., Sp. Publ. (7):71-93, 1970.
30. Office of International Fisheries, Norwegian salmon and trout farming. U.S. Dept. Comm., Natl. Mar. Fish. Serv., *F/M321:GH/DW, IFR:84/10* (1984).
31. D. J. Edwards, "Salmon and Trout Farming in Norway." Fishing News Books Ltd., Farnham, Surrey, England, 1978.
32. Office of International Fisheries, World salmon farming. U.S. Dept. Comm., Natl. Mar. Fish. Serv., *F/M321:BD. IFR-84/89B-R* (1984).
33. D. A. Conroy, An evaluation of the present state of world trade in ornamental fish. *FAO Fish. Tech. Pap.*, (146):1-128 (1975).
34. H. R. Axelrod, The aquarium fish industry--1971. *Tropical Fish Hobbyist*, pp.19-23 (Nov. 1971).

Chapter 8

The Anglers' Edge

Anglers, by their numbers, their sturdy adherence to the ancient right of piscary, their ideal of conservation, their expenditures, and their sparing use of fish, have a significant influence on government fishery policies for all public waters in which they fish and for all stocks that they favor. This is true in all of the fresh and coastal waters of Europe and northern America, and on the high-seas off both continents for such prizes as billfishes and tunas. The anglers' edge is to have fish of the commons allocated to them by getting together to request the allocation and paying for the required government services.

Angling was probably a sport of kings before horse racing, when those who could order someone to fish for them chose instead to fish themselves for the sport of it. Drawings in Egyptian tombs dated more than 3400 years ago show leaders engaged in fishing, and tales of fishing in the Mediterranean region 2000 to 3000 years ago give accounts of angling methods.(1)

Fishing has long played a major role in the lives of coastal and riparian peoples--in both practical and mystical ways. Early fishermen in the Middle East worked as "servants of the gods"; the temples managed the fishing and accepted gifts of fish to the gods.(2) At that time, and still to this day, some anglers catch fish for family food in easy and practical ways. But angling has elements of mysticism, as people accept challenges to catch fish in the most difficult way possible: with fragile equipment, with tiny lures such as dry flies, and with lines or leaders having a strength of only a small fraction of the weight of the fish. They find a personal satisfaction unique to their own challenges and experiences in hooking, playing, and landing fish under such conditions. As one angler has expressed it,

> Fortunately, I learned long ago that although fish do make a difference--*the difference*--in angling, catching them does not; so that he who is content to not-catch fish in the most skillfull and refined manner, utilizing the best equipment and technique, will have his time and attention free for the accumulation of a thousand experiences, the memory of which will remain for his enjoyment long after any recollection of fish would have faded.(3)

This chapter is about angling, which is the principal recreational fishing activity, but also included are shellfish gathering and fishing with gear other than rod and reel, where legally used and when the catch is not sold.

8.1 A SOCIAL PREFERENCE

The experience of fishing for, catching, or not catching fish is treasured by a rapidly increasing proportion of people in industrialized countries who have time and income to enjoy the sport. The activity has evolved from essential subsistence fishing and most anglers still value the food obtained, but getting away from regular routines, relaxation, companionship with friends and family, and enjoyment of clean natural surroundings are the major attractions of angling.

Recreational fisheries have been developing rapidly since 1950. The statistics are best, but inconsistent and fragmentary, from northern America and European countries where anglers are a large proportion of the population. Angling is also increasing rapidly in less developed countries near cities and wherever tourists gather. It has become a major outdoor recreational activity in many parts of the world.

Angling is a sport for about 25% of the U.S. population whose total catch is estimated to equal about half of the annual commercial harvest of edible finfish. About 54 million anglers were estimated to have caught about 0.85 million tons of fish worth about US$2.5 billion in 1975. Their numbers have been increasing at a compound annual rate of more than 3%, about twice the rate of population growth.[4]

A similar percentage of the population and a large number of nonresident tourists go angling in Canada. About 4.9 million residents and about 1.2 million tourists went angling in 1975, and their numbers had been increasing at about 4% per year. They caught about 8% of the combined catches of commercial and recreational fisheries but paid nearly six times as much in license fees as the commercial fishers. Their total direct expenditures of about C$1.1 billion was greater than the total market value of the commercial fisheries--C$694 million.[5]

Participation in recreational fishing in Europe varies widely from country to country (the accuracy of the statistics is variable) but the average appears to have been only about 3.7% of the population in the early 1970s. Still, it involved more than 14 million people then, with the highest percentages of the populations in the Scandinavian countries, France, The Netherlands, and the United Kingdom. The degree of participation appears to be related inversely to the extensive water pollution in central Europe and directly to the numbers of salmon and trout streams. The numbers had been rapidly increasing in several countries which reported doubling of anglers in the decade of the 1960s, and a projection to 1980 indicated that the total for Europe might then have been more than 50 million.[6] Japan is another country with rapidly growing numbers of anglers.

One of the reasons for the popularity of angling is the diversity of motivation and activity. The patterns have been listed by William Dill as follows:[7]

1. Pleasure incidental to the capture of food
2. A pastime gradually codified as sport and extended from nobility to all people
3. Recreation coupled with enjoyment of nature, good fellowship, and simple pleasures
4. Adventure in distant, virgin, or brawling waters
5. Competition for prizes, trophies, or money
6. A basis for business, tourism, real estate, supplies, equipment, lodging, services, etc.
7. Relief of tensions, promotion of health and well-being

Such diversity provides a singular advantage in the development of angling. The central issue in almost all waters where angling takes place is integration of angling with other uses of the waters. Every public water is used for something: commercial fishing, boating, rafting, water skiing, sailing, water supply, power generation, navigation, waste disposal, swimming, and perhaps a dozen others. The problem that follows as anglers protect or expand their activities is the resolution of conflicts with the multitude of other uses. The anglers can be flexible in enjoying a sport: bait fishing from a city pier, teaching a child to fish, fly fishing in a mountain stream, competing for the biggest fish taken in an afternoon, or trying to catch a salmon with a million dollar tag that was released as a business advertisement. An angler can almost always fish for something of personal value except in places desecrated by pollution.

Anglers have commonly accepted the principle that taxes levied on the general public should not be used to support a special recreation for only certain people. In the United States anglers pay directly for most of the government services to angling and anglers. State management and enhancement programs are largely paid for by license fees, and national research programs are supported by special taxes on angling equipment; this in contrast to very small license fees and catch taxes levied on commercial fishers that pay little of the public costs.

Anglers in northern America and in most European countries have become extensively organized into local clubs, state federations, national federations, coalitions, and associations; the pattern and functions vary in each country. In European countries, they may have licensing and management responsibilities.[6] Some anglers' organizations are public, but most are semi-private or private.

Yugoslavia, for example, in the early 1970s had 493 associations, 41 clubs, 8 regional unions, and a federal union with management responsibilities. France had about 2.3 million anglers, 4000 angling clubs, and 92 angling federations in 1984. According to an investigation in 1976, Sweden had about 1.4 million anglers, about 350,000 of whom were owners of fishing waters, and about 670 clubs. The Polish Anglers' Association with about 400,000 members, stocks fish, collects statistics on catch, builds facilities, and cooperates with the state tourist agency. Anglers must belong to the Association. Spain has a system of "cotos" (organizations with authority to manage stream sections) to which anglers must belong. There were 523 in 1971, mostly for salmon and trout, with full management responsibilities. In addition, 23 countries in Europe and Latin America are associated with the Confédération Internationale de la Pêche Sportive (CIPS), which actively fosters international development of sport fishing.[6]

Canada and the United States have thousands of local clubs, hundreds of provincial or state federations, and dozens of national federations. A small proportion is concerned exclusively with angling; the majority with conservation, wildlife, and the environment in general. Some major national or international federations with networks of provincial or state and local branches are the Canadian Wildlife Federation (457,000 members), the (U.S.) National Wildlife Federation(4.1 million members), the Isaak Walton League of America, Inc. (50,000 members), Trout Unlimited (49,000 members in 1986), the Federation of Fly Fishers (8600 members), and the Bass Angler's Sportsman Society. Other influential private angling organizations are the International Game Fish Association, the National Coalition for Marine Conservation, and the Sport Fishing Institute.[8]

All of these organizations provide information services to their members, especially about government policy and programs affecting the fishing. They all work closely with provincial, state and federal governments to advocate research and to influence legislation and regulations. At local levels the members participate extensively in fishery management activities, such as fish tagging, fish stocking, and environmental improvement. At national levels they have played a major role in changing the objective of fishery management, that had been stated in treaties and laws of the 1940s and 1950s, of obtaining the maximum sustainable yield. This was modified in the early 1970s during the Third Law of the Sea Conference with the added phrase, "taking into account relevant social and economic factors". At a U.S. conference in 1974, a definition of *optimum yield* was advanced as "a deliberate melding of biological, economic, social, and political values designed to produce the maximum benefit to society from a given stock of fish."(9) Such a definition allows consideration of value judgments, such as nonextractive uses and quality of the angling, in addition to overall economic benefits. More explicitly, it recognizes the reality of changing political concepts.

The political strength of anglers lies in their numbers and organizations. In the United States they dominate the management of freshwater fisheries and, for the past decade, have been rapidly increasing their influence on management of the saltwater fisheries. One reason for their success is that anglers outnumber commercial fishers by about 300 to 1. About 36 million licenses, tags, permits, and stamps were sold in fiscal year 1984 at a cost to fishers of US$260 million--and substantial numbers of saltwater fishers, elderly people, and young people were exempt from license requirements.(10)

8.1.1 Support of Environmental Quality

These large fractions of the population are major political forces in favor of clean and fishable waters. The quality of the outdoor experience, as well as the opportunity to catch fish, depends on good water.

Some of the functions of the anglers' organizations are to sound alarms about deteriorating water quality, to identify opportunities for rehabilitation of damaged waters, and to advocate attention to fishing values in water development projects. Typically they advocate good science and support projects based on sound information.

8.2 THE BUSINESS IN SUPPORT OF ANGLING

Anglers constitute a large fraction of the tourists who visit shoreside resorts; indeed some resorts cater almost entirely to anglers. They contribute, therefore, to a wide variety of businesses supplying transportation, lodging, food, beverages, boats, tackle, fuel, and bait at retail, as well as at wholesale, manufacturer, or producer levels.

In England and Wales, for example, a 1972 survey produced an estimate of 2.8 million anglers who spent an average of about £80/year each, or gross expenditures of £200 million to £250 million, about two thirds of which related to inland waters. Sales rights to fish for salmon during the preceding five years, in which prices were based on average catches in the streams, ranged from £500 to £750 per fish caught! Sport fishing for some species such as salmon has become so valuable that political pressures increase to restrict commercial trolling and net fishing.[6]

In the United States a 1980 study found that direct expenditures for goods and services associated with marine recreational fishing were approximately US$4 billion, about double similar data for 1975,[11] whereas the total value of the U.S. commercial catch as landed in 1980 was US$2.3 billion. By comparison, the amount spent for freshwater recreational fishing was about twice as large as that spent for marine recreational fishing.

The immediate expenditures of anglers also generate much additional economic impact; the "multiplier effect" is usually an additional 50-100%, especially for private and community investments. In the 1980 study, the total economic impact of the U.S. marine recreational fisheries was estimated as about US$7.5 billion. This was estimated to have involved about 65,000 person-years of employment to provide the equipment and services.

In addition to the monetary values, the social values of angling may be considered. They are associated with clean water, and in The Netherlands, for example, angling development is an integral part of development planning for reclaimed areas. Plans for pollution abatement are frequently based on the water quality needed for the species of fish desired. Angling and clean water are strong tourist attractions in Europe, especially for the countries with trout and salmon waters. The northern countries attract millions of anglers from neighboring countries, and Yugoslavia reports about equal numbers of tourist and Yugoslav anglers.[6]

8.3 DEVELOPMENT ACTION

Anglers have acquired an identity and have developed an almost universal ethic. When they depend on the commons, they use it sensibly. They have gathered economic and political strength in an increasing number of countries to form a community which seems certain to endure and grow. Undoubtedly, this community will increase its influence on the government agencies managing the waters and fish with which they are concerned.

Development of angling in the public waters requires political coordination of the angling constituency. The action should focus on achieving public recognition of the activity--its size and economic importance, its diversity with respect to participation and supporting businesses, and its fundamental concern with wise use of the resources of the commons. This can be done through work by anglers' organizations with legislatures, Chambers of Commerce, tourism and travel groups, and investors concerned with recreational activities. Following that, the action can identify abuses of environment and fish resources and press government agencies to investigate and correct such problems.[12]

In contrast to the development of market fishing (see Chapter 10), the policy issues of angling development are of much more concern to regional and national governments than to local governments. Catering to anglers adds economic benefits to local businesses, while seldom causing a loss of livelihoods. Even in coastal marine communities, when a species has been allocated to anglers, the market fishers for other species can usually share in the overall community benefits.

NOTES AND REFERENCES

1. W. Radcliffe, "Fishing from the Earliest Times." John Murray, London, 478 pp., 1921.
2. R. Kreuzer, The cradle of sea fisheries. *In* E. M. Borgese and N. Ginsburg, eds. "Ocean Yearbook 1." Univ. of Chicago Press, Chicago, pp. 102-113, 1978.
3. Sparse Grey Hackle (Alfred W. Miller). "Fishless Days, Angling Nights." Crown Publ., New York, 223 pp., 1971. Quote from p. xiii.
4. R. F. Hutton, P. L. Hooper, and R. H. Stroud, United States of America country review. *In* J. H. Grover, ed., "Allocation of Fishery Resources: Proceedings of the Technical Consultation on Allocation of Fishery Resources, Vichy France, 20-23 April 1980." FAO and Amer. Fish. Soc., pp. 613-623, 1982.
5. A. L. W. Tuomi, Canada country review. *In* J. H. Grover, ed. "Allocation of Fishery Resources: Proceedings of the Technical Consultation on Allocation of Fishery Resources, Vichy France, 20-23 April 1980." FAO and Amer. Fish. Soc., pp. 539-544, 1982.
6. J.-L. Gaudet, Economic issues and opportunities facing Europe in the field of sport fisheries. *EIFAC Occasional Pap.*, FAO, (7):1-14 (1973).
   -P. Affre, Fishing in southern Europe, especially France. *In* R. H. Stroud, Ed., "World Angling Resources and Challenges." Int. Game Fish Assoc., Fort Lauderdale, FL, pp. 61-66, 1985.
   -O. Bergfors, Angling in Sweden. *In* R. H. Stroud, Ed., "World Angling Resources and Challenges." Int. Game Fish Assoc., Fort Lauderdale, Florida, pp. 67-69, 1985.

7. W. A. Dill, Patterns of change in recreational fisheries: Their determinants. *In* J. S. Alabaster, ed., "Recreational Freshwater Fisheries: Their Conservation Management and Development: Proceedings of a Conference of the Water Research Centre, Keble College, Oxford, 12-15 Dec. 1977. Stevenage Lab., Elder Way, Stevenage, Herts., England, pp. 1-22, 1978.
8. National Wildlife Federation, "Conservation Directory, 1984," 30th Ed. Nat. Wildlife Fed. Washington, D.C., 279 pp., 1984.
9. P. M. Roedel, A summary and critique of the symposium on optimum sustainable yield. *In* P. M. Roedel, ed. "Optimum Sustainable Yield as a Concept in Fisheries Management." *Amer. Fish. Soc., Spec. Publ.* (9), pp. 79-89, 1975.
10. Sport Fishing Institute, Number of paid fishing license holders, license sales, and cost to fishermen, fiscal year, 1984. *SFI Bull.* (367):4 (Aug. 1985).
11. Sport Fishing Institute. Marine recreational fishing--big business. *SFI Bull.* (348):1-2 (Sept. 1983).
12. C. R. Sullivan, Recommendations for action: Panel 2, the marine recreational fishing industry. *In* R. H. Stroud, ed., "Marine Recreational Fisheries 9," Nat. Coalition Mar. Conserv. Inc., Savannah, Georgia, pp. 193-195, 1984. See also several other papers in the volume.

PART III

FUTURE STRATEGY

The goals of fishery development are enticingly simple: more food, recreation, employment, and income, especially for poor people. What could be more obviously beneficial, with fish so widely distributed among inland, coastal, and ocean waters that cover more than two-thirds of the earth?

Fishery development has been a continuing evolutionary activity, but now most governments, stimulated by recently extended authority over coastal zones, want to accelerate the process. The problems that have been encountered seldom yield to simple or theoretical solutions; they require the leavening of experience in an art rather than a science.

The problems of market fishery development are discouragingly complex. We have already increased total production from most market fisheries of the commons as much as we can. Further development, in the long term, must involve effective government management for conservation as well as reallocation among users. The short-term approach, based on universal rights of piscary, is an economic and political trap. More fishing, better equipment, and more employment lead only briefly to more production and disrupt the livelihoods of traditional fishers. Then the fishing becomes a business with too little employment, too much debt, too little income, and need for government help. The government helps with subsidies and compounds the trouble. The fishing becomes an economic drain and a social disaster.

Few governments have avoided the trap, but it can be avoided with policies that are developed in close collaboration with the people affected, once they recognize the trap. The challenge is to focus on the problems and perceptions of the fishers and farmers in order to find the blend of technical, policy, and social changes that will help them.

This section is an attempt at a comprehensive approach to an activity that typically has been approached from narrow viewpoints. The objective is to stimulate an appreciation of the overall problems that demand the best efforts from groups of specialists.

Chapter 9

Constraints on Fishery Development

What does a carp and duck farmer in China have in common with a fly fisher on a private stream in Scotland, an aquarist in Rotterdam, a master of a thousand-ton trawler-processor in the Bering Sea, a Canadian housewife shopping for frozen cod fillets, or a hungry spear fisher adding to a family's diet of jungle food? Not much--except that all will be influenced by their country's fishery policies. Such a diversity of human interests in the fisheries is seldom recognized in government fishery policies, university fishery schools, or by the public. Add to it inherently rapid rates of change in angling, fish farming, market fishing, fish products, fish markets, and government controls and find an explanation for widespread public confusion over fishery issues.

This may explain why no simple concept of fishery development can be described or even exist. But fishery development is a common objective of all national governments and many state or local governments. The impressive gains in fish production of the 1950s and 1960s, the EEZ defined in the new LOS, and the international publicity given to food from the sea have created a new awareness of the fisheries.

No manual on fishery development has been written, nor is this an attempt at one; the topic is too complex and too little understood. But fishery development projects are underway in most countries, and their success or lack of it has provided a rich background of experience. A few examples have been recounted in earlier chapters, and these plus the published record of many others provides a basis for indentifying some principles.

The overall goals of fishery development are deceptively simple: use of the fisheries to increase economic benefits while maintaining social equity. It is not only impossible to maximize these goals simultaneously, but they conflict in ways that vary with time and kind of fishery, kind of markets, and changes in the resources. Discovery and adjustment of the compromise between them is the constant political challenge.

Most fishery development projects have focused on technical and economic constraints to the neglect of pervasive policy constraints by governments. All three kinds affect fish farming, angling, and market fishing but in very different ways and to very different degrees. This chapter will consider first the simpler development problems of fish farming and angling and then discuss at some length the relatively intractable problems of market fishing development.

9.1 CONSTRAINTS ON FISH FARMING

Fish farm development, to the extent that it does not rely on the commons, is similar in many respects to agricultural development. The two are technically linked through similarity of the forage systems (see Section 7.3) in which fish are grown along with other farm crops; they are socially linked through similar rural populations and communities with few opportunities for alternative employment. The more sophisticated kinds of animal production are linked through similarity of the feed systems in which fish, poultry, or mammals are grown in factory-like systems (see Section 7.4). And administrative linkage occurs in many governments through service to fish farming by departments of agriculture.

Governments can assist with such development through projects that provide research, extension services, infrastructure, credit, and education as long as they avoid stifling measures such as price and trade controls that have discouraged farmers. Accordingly, development of fish farming has been addressed by an FAO study group as essentially comparable to agriculture. The principal controlling factors were identified as the environment and space suitable for the fish, the technology of farms and products, the management of the businesses of production, and the management of the businesses of marketing.(1)

When the fish farmers depend on the commons, however, for brood stock, seed, or farm sites, they face political constraints in addition to the economic and production constraints on the supply of those items. Such constraints may arise from the active hostility of the fishers or through a need for allocating the supply between fishers and farmers.

Even without the problems of the commons, government assistance is a challenging task. Often the people are without experience in fish farming and encounter major managerial or social difficulties with new ventures (see Chapter 10).

9.2 CONSTRAINTS ON ANGLING

Almost all anglers use the commons; the few exceptions are mostly private clubs or fee-fishing ventures that own fishing rights which are provided in addition to social benefits. Yet the development of angling, at least for the next few decades, appears to be relatively independent of the constraints related to use of the commons that apply to market fishing. The government policies may impact the pleasure of the fisher and the economy of the supporting services, but not the livelihood of the fisher.

Anglers can accept increasing restrictions on catch for the purpose of conservation as long as the amenities of the experience are preserved. The value of the privilege usually increases as the opportunity to catch a fish declines and the economic benefit continues to grow. Indeed, the competitive aspects of catching a rare fish provide a treasured experience. The anglers have thrived as commercial fishing gear has been restricted or banned in the interests of conservation. They are so numerous in the developed countries that they can obtain allocation of fish resources for their exclusive use. And usually they pay for a large part of the direct government services by acceptance of license fees and special taxes on their activity (see Chapter 8).

A government view of angling development in the United States stressed the importance of keeping the fishing good through maintenance and improvement of the habitat, through stocking programs, through information programs to describe angling opportunities and encourage use of abundant species, and through improvement of the infrastructure relative to angling areas.(2) Particular measures include enhancement of fish production in new lakes, development of artificial reefs by sinking old ships or oil rigs, and providing easy public access to more fishing waters.(3)

Such government programs start when anglers get the attention of governments. They have done this in Australia through the recent formation of the Australian Recreational Fishing Confederation that brought together seven relevant national associations in order to promote environmental improvement, fishery research, management, and better enforcement.(4)

The primary constraint on development of angling appears to be maintenance of the clean and pleasant aquatic environment that is essential for the fish as well as for all water-based outdoor recreation.

9.3 CONSTRAINTS ON MARKET FISHING

Most of the efforts to develop fisheries have focused on production of food rather than fun. This emphasis is, of course, consistent with efforts to feed hungry people rather than augment tourism. But it has also followed an optimistic euphoria about the potential for increased production from marine fishing generated by three decades of rapid growth and perceptions of benefits from extension of coastal jurisdiction to 200 miles.

9.3.1 The Fish Stocks and Their Variability

The world's fish production has slowed from its 6-7% annual rate of increase of the 1950s and 1960s to between 1% and 2%, and it is likely that sustaining even this rate will require better management for conservation (see Section 4.2 and Ref. 4-9). The necessary improvement in management is unlikely to be accomplished because of unpredictable variation in most stocks that makes effective management for conservation nearly impossible, and because of ineffective management organizations even in the most advanced countries.

The scientific basis for stock management with the objectives of obtaining either optimum sustainable physical yields or optimum sustainable economic yields has turned out to be inadequate because of extreme and largely unpredictable variability in the size of the stocks. This inadequacy results in poor recommendations for regulation which, in turn, generate political distrust of the regulatory agency and widespread evasion of laws.

One kind of variability, fortunately rare, occurs in stocks of very slow-growing species that may become quite abundant. These may sustain a large fishery for several years but, if overfished, which is often easy after discovery, the stock may require complete protection of the spawners for many years in order to recover.

The variability of major concern, however, is the interannual variability associated primarily with the size of year classes (cohorts) entering the fishery and their rate of decline. Each cohort in most species with average growth rates contributes significantly to catches for five years or less; usually dominating the catches for only a year or two. The other common variability in the catches, that associated with changes in the seasons, is caused by migrations around feeding and breeding times and is well known to fishers and scientists.

The scale of the variability in stock size varies according to the fecundity and habits of the species, but it is so large for almost all oceanic stocks that any fisher, dependent on a heavily fished stock of a single species, will probably have an economic failure at least one or two years out of every decade.

The largest variability appears to be in the pelagic stocks, such as herrings, anchovies, pilchards, and mackerels, which may vary in stock size by factors of more than 100. Consider, for example, the decline in herring abundance in the eastern Bering Sea from about 1.7 million tons in 1963 to almost none in 1980, with an overlapping increase in the Alaskan pollock (semipelagic) from almost none in 1962 to about 17 million tons in 1972.[5] Or the decline in the California pilchard from about 3 million tons in 1932 to only a few thousand tons in the early 1960s, with an overlapping increase of the anchovy from a few thousand tons in 1955 to about 2 million tons in 1975.[6] Changes of similar magnitude, frequently with one species replacing another, have occurred in the herring and mackerel fisheries on both sides of the North Atlantic; in the pilchard and mackerel fisheries off Japan; and in the pilchard, jack mackerel, and anchovy fisheries off Southwest Africa as well as off Chile and Peru.[7] Many such stocks are large enough to sustain annual catches of 1 million tons or more during periods of abundance.

Most of the bottom fish stocks such as members of the cod and flounder families are smaller and much less variable but still fluctuate enough to require frequent changes in management regulations and the fisher's strategies. For example, 15 stocks in the Gulf of Maine have commonly shown fourfold variability from 1/2 to 2 times the median size, but a few have shown much greater changes such as the important Georges Bank haddock from 1/10 to 3 times the median size.[8]

The causes of the fluctuations are certainly complex but poorly understood. They arise from the dual life of almost all marine species in which eggs and larvae drift helplessly with the currents, when they are part of the plankton and far more vulnerable to predation, competition, and the vagaries of the currents than they are later as foragers and active participants in the communities. During their drifting life they are subject to the abiotic factors of currents, temperature, salinity, and dissolved oxygen, as well as to the biotic factors of food supply, predators, and competitors. Included in the biological definition of predators are epidemic diseases, fortunately rare, but which occasionally decimate populations or make the flesh inedible.

The mortality rates from the combination of factors during the drifting life can be extremely high. An investigation of Atlantic mackerel spawning in 1932 found that the mortality rate of eggs or larvae was as high as *30-45%/day*, even though the larvae at one stage grew at a rate of *12.5%/day* in weight. The result in this example was a failure of a year class that was probably due to unusual currents moving the larvae away from their accustomed nursery areas. Only between 1 and 10 mackerel reached a length of 50 mm out of each million eggs spawned, far too few to be likely to provide two adults for the next generation.(9)

Obviously, such rates can be sustained only for very brief periods, and a difference of a few days exposure to critical conditions can make a major difference in abundance of a year class. Equally obviously, monitoring and understanding such complex short-term events is a difficult challenge for scientists. The time and space scales covered by events potentially involved with large oceanic populations are so great that even identifying the crucial factors is difficult, and routinely measuring their relationship to abundance of fish stocks has been generally unsuccessful.(10)

Even in the North Sea, where the relationship of oceanic conditions and fish stocks has been studied longer and more thoroughly than anywhere else, the relationship between abundance of recruits and abundance of parent stock as modified by ocean conditions, remains virtually unpredictable. The existing monitoring programs have added little to understanding the mechanisms behing the changes. Furthermore, a secondary effect on the fishery management system is loss of confidence in it by the fishers and widespread falsification of catch records.(11) Optimizing the net economic yield of such fisheries by refinement of the monitoring and regulatory system appears to require government costs greater than any possible benefits.(12)

The forecasts of abundance, which must be made for all optimum regulatory systems, should therefore be based on cohort abundance after a cohort has appeared in the fishery. Current forecasts based on the relationship between spawners and progeny are frequently misleading for either government planning of regulation or for business planning of fishing and marketing. Furthermore, all biological and economic theories, based on average expectations from a steady-state condition of single species, are of little value to either business people or governmental managers.

In addition to the resource variability, the market fishers must expect variability in markets caused by changes in competing products, some of which may be traded internationally, and variability in costs, especially fuel. Few studies of the combined impact on individual fishers have been made, but a review of eight major U.S. fisheries, comprising about 46% of the total value of U.S. landings, provides evidence of fishery-wide problems.(13)

The study used 1965 as a base year, with an index value of 100, and covers the 18 years through 1982. A "health index" was computed for each year for each fishery based on time spent fishing, catch per unit of effort, costs of fishing operations, and prices received. The final index values of the health of eight U.S. fleets showed declines in six, no change in one, and only a 9% increase in one (Table 9-1).

These indices at the end of the period mask surprising year-to-year variation in the situation of the fleets (Table 9-2). Five fleets had ratios of the best to poorest year greater than two, and the most stable, Maine lobstering, had a ratio of 1.36. Year-to-year ratios in most fleets commonly differed more than 10%. Given industrywide changes of this magnitude, large numbers of individual fishers must have experienced much greater variability in income and return on investment.

Furthermore, during the 18-year period, the consumer price index for fish and shellfish rose fairly steadily from the base of 100 in 1965 to 408 in 1982, a substantially greater increase than that for beef and veal (293), or poultry (193). Thus, such indifferent performance for this large segment of the overall fishing industry was related primarily to internal factors rather than to any overall decline in relative fish prices.

Sustaining a livelihood in the face of such variability has long been a problem for most fishing businesses.

Table 9-1.
Indices of U.S. industry health in 1982--base of 100 in 1965.

New England scallop dredging	0.98
New England otter trawling	1.00
Maine lobstering	0.90
Middle Atlantic surf clam dredging	0.83
Gulf of Mexico shrimping	1.09
Gulf of Mexico menhaden seining	0.72
Pacific tuna fishing (except albacore)	0.60
Alaskan king and tanner crab fishing	0.94

Table 9-2.
Ratios of best to poorest year indices, years 1965-1982.

New England scallop dredging	2.62
New England otter trawling	1.38
Maine lobstering	1.36
Middle Atlantic surf clam dredging	2.11
Gulf of Mexico shrimping	1.94
Gulf of Mexico menhaden seining	4.08
Pacific tuna fishing (except albacore)	2.23
Alaskan king and tanner crab fishing	2.38

They manage by shifting the fishing from stock to stock according to abundance and market conditions. The fishers have developed gear adaptations where necessary for the different species and learned about their different habits. The processors and marketers also have been accustomed to change quickly from one species to another, seasonally as well as interannually.

After coping with the problems of the sea and the fish stocks, every fisher has had to be optimistic to continue fishing. When government regulations restrict fishing on the basis of the questionable accuracy of the stock-recruitment relationship, fishers naturally resist and are joined by politicians who will be concerned with related problems. During the crisis period of the Peruvian fishery, for example (see Section 6.4), the scientists recommended closures of the fishing, but the government, faced with a financial crisis and already subsidizing the fishing industry, chose to continue fishing with the result that the anchoveta stock was almost depleted. Few fisheries are of such national importance as the Peruvian anchoveta, but numerous regional or local economic problems will always create pressure for a more optimistic approach.(14) Even ideological perspectives were invoked, such as an assertion that the Peruvian anchoveta fishery led to the superexploitation of the proletariat.(15)

Furthermore, with the rapid expansion of international fish markets and transnational fish corporations, international pressures may arise. The Peruvian anchoveta collapse naturally affected the world market for fish meal and led businesses in other countries to consider purchase of idle vessels and processing plants or otherwise expand their fish meal production with an expectation of rising prices.(16)

9.3.2 Government Perceptions

The most comprehensive and pragmatic examination to date of the fishery development process was made at a meeting called by the FAO in 1983, which was intended to provide preliminary information for its 1984 Conference on Fishery Management and Development. Case studies were requested from 11 countries, 3 each from Asia, Africa, and Latin America plus 2 from Europe, in an effort to obtain representative experience. The case studies plus 13 additional interpretative papers discussed development only of marine food fishing, although the current role of fish farming was described.

The countries included Brazil, Cuba, Indonesia, Japan, Norway, Peru, Poland, Senegal, Somalia, Sri Lanka, and Tanzania; the results were intended to reflect the ecological, socioeconomic, and political situations in different continents. Included were countries with highly developed and less developed economies and fisheries, with distant water and nearby fishing, and with new and old governments. It appears to be justified, therefore, to use the results as representative of a large group of countries.

The conference summary revealed the diversity of views.(17) There was no complete agreement among the 11 countries on any single objective, constraint, or strategy for fishery development. The closest approaches to unanimity appeared in the desired objectives with 9 countries agreeing on increasing fish consumption, 8 on improving foreign exchange, 6 on increasing fish production, and 5 on improving living conditions of fishers. Among the total of 26 constraints, those most frequently mentioned were an increase in fuel and operational costs (7 times), inadequate information on fish resources, lack of appropriate fisheries development institutions, lack of appropriate marketing infrastructure, and isolation and lack of services in fishers communities (5 times each). Among the 19 elements for alternative strategies there was similar diversity; improvement of inland fishing and fish farming was mentioned by 6 countries, as was special assistance to small-scale fisheries. Most of the rest were mentioned 3 times or less.

Only 3 of the 11 countries had specific fishery development plans; 6 others included fisheries in their national development plans. Little attention was given to policy formation and implementation, to potential conflicts among policies, to the need for continuing review and evolution of policies, or even to counterproductive and inefficient policies. Some of the countries, including Peru (see Section 6.4), mentioned the need for new policies, but none looked critically at past policies that had failed, at the legal and institutional instruments of policy, or at the variability of the resources. Only Japan emphasized the fact that its fishery development was the result of a long historical process that included a progressive change in fishery policies to fit the needs of each development stage (see Section 6.6 and Ref. 6-48).

The diverse views are not surprising, because few if any human activities considered in a general category such as fisheries are as diverse technically, socially, and politically. They engage some of the largest transnational corporations in the world as well as individual anglers or subsistence fishers, include hundreds of species and thousands of products, make use of private farm facilities and public resources of the commons, and challenge national governments to fit laws and international agreements into the accustomed practice of a rural village. On top of all this, yesterday's policies may not fit today's realities.

The corporate-individual conflict appears in most countries between the large-scale and small-scale or artisanal fishers, who are frequently pursuing the same resources and selling to the same nearby markets. The large-scale operations may also enjoy major economies of scale by selling standard products to national or international markets after becoming known in those markets. The small-scale fishers, on the other hand, are usually following a traditional life, shaped over decades or longer, in pursuit of a few local species and lack resources or abilities to change their life-style (see Sections 6.1, 6.3, and 6.5).

Fish are so diverse to the consumers, almost all of whom have experienced products of poor quality, that a change of customary marketing systems is difficult, especially with fresh and cured products. Most of these are sold on the basis of the reputation of the local retailer, not on the basis of a brand name that has earned widespread acceptance in distant markets. International marketing requires market knowledge and special skills developed over a long time (see Section 6.7 about the troubles of the United States in developing foreign markets).

This set of papers[17] leaves the impression of government preoccupation with problems of the moment, especially with the kind of financial and technical issues that had overtaken the marine market fisheries at the end of their expansion era. They fail to consider the much more difficult government role that has become necessary for fishery management as well as development.

9.4 THE POLICY SHORTFALL IN MARKET FISHING

It seems obvious that, with a critical relationship between fishery policies and successful fishery development, every fishery development project can succeed only if it is supported by compatible government policies over a long term. In fact, a lack of compatibility is probably second only to the supply of wild fish as the major constraint on fishery development in the long term. The reasons for this deserve elaboration.

The special difficulty for governments is managing the resources of the commons, of which both fish and water are typical. If we are to avoid destruction of the resources and achieve equity in their development, their usage must be dealt with by one or more of the three major political levels: the national level of sovereignty, including the intermediate state or provincial divisions; the local governments or organizations; and the international relationships that must be considered in most marine and major river fisheries. The sharing of responsibility for the commons among the levels of government is exceedingly complex for long-established nations and especially fragile for newly established nations.

9.4.1 National Sovereignty

The fishery management practices by national governments, or the intermediate-level state or provincial governments, have dominated the government fishery activities in most countries. One level or the other has taken over most fishery management activities, because a majority of the fishery resources migrate among or span local jurisdictions and require a uniform system of management. The national and intermediate levels have also amassed technical expertise to manage the resources. The division of responsibility between national and intermediate levels is usually defined in law and is relatively well established.

The nearly universal approach to fishery management has become based on the concept of obtaining a maximum or optimum yield as determined through scientific study of the resources. This has emerged worldwide only after the scientific discoveries of the first half of the twentieth century and the adoption of national or international laws since 1950; in fact the fishery part of the LOS has been widely recognized for only about a decade. This part of the LOS has been based on broad acceptance of the use of science to approach the objectives of conservation and the maintenance of natural resources.

The relatively sophisticated scientific studies have been widely adopted by governments as a basis for estimating the status of the fishery resources, predicting the catches possible with different amounts of fishing effort; and regulating the fishing with respect to time, place, gear, and catch. The system recognizes the ancient rights of piscary and usually tries to maintain the ancient competitive relationships among fishers as basic principles. When the regulations are widely accepted by the fishers and accompanied by effective enforcement, the objective of conservation can be achieved. But when major social stresses occur, as between recreational and commercial fishers or between small-scale and large-scale operators, the regulations may be widely evaded.

Clearly the scientific concepts are too simplistic and too narrowly focused.[18] One shortcoming is the difficulty of predicting major failures in some large stocks with evidence convincing to the fish businesses (see Section 9.3.1). Such collapses have occurred erratically, often after vigorous and profitable fishing has been pursued for a decade or more, and have not been scientifically predictable far enough in advance to guide investment decisions.

Another serious shortcoming has been the inability of scientists to predict the economic and social impacts of those regulations intended primarily for conservation. These vary greatly among localities and even among those dependent on the same resource, according to local conditions of access and alternative opportunities for employment. Even if anticipated, the different local problems can seldom be dealt with satisfactorily by blanket national or regional approaches. The result in such cases has been social problems in many localities where the national and intermediate level governments have failed (see the discussion in Section 6.5 of the social disaster in Eastern Canada).

These shortcomings have occurred in spite of large continuing fishery research programs by dedicated scientists. The Canadian fishery research, for example, is widely regarded by fishery scientists as exemplary, yet its nationally regulated fisheries face major social problems.

The most acute social problems among local fishers, however, appear to be associated with introduction of improved fishing technology in addition to regulation for conservation. Such introductions are the normal sequence of events in the lives of fishers who always want better gear, and they are also the usual result of national development efforts which focus on technical assistance to the fishers.

Clearly, when the resources become well known, efficient new vessels can be designed for catching a species or group of species in a known fishing area with a maximum of efficiency. In less developed fisheries it is entirely possible for such new vessels to catch 10 times or even 30 times as much fish *per person* as the older multipurpose equipment and methods. If the yield of the resource is limited by law, or near its maximum yield in the absence of law, the inevitable result will eventually be displacement of 9 or even 29 fishers from their occupation (see Table 3.2 and Section 6.5). Even if the yield from the resource can be expanded, the easy addition of such an increase in fishing effectiveness can quickly bring the yield above the maximum sustainable, thereby producing both an economic and conservation disaster.

On the other hand, such an increase in efficiency may be necessary to remain competitive in the market. Fish must not only compete with other protein foods, but they must remain competitive in the fish markets which now require national and international product identification and marketing skills.

The plight of the U.S. fisheries is an excellent example of the course of national and intermediate level management (see Section 6-7). The Fishery Conservation and Management Act of 1976 followed the principles of national marine fishery conservation reached by consensus at the Third LOS Conference. It was based on a shared authority--the fisheries inside 3 miles remained under authority of the states, while those between 3 and 200 miles were under authority of the federal government. The state and federal governments were required to cooperate in a system of 8 regional councils. The councils were given authority to originate annual plans for fishery management after consideration of the conservation research being done by the states and an expanded federal research program. Final approval of plans, however, remained with the federal government, so it was really a centralization of authority.

The experience of U.S. New England fishers following passage of the FCMA in 1976 provides an example. At first the fishers were enthusiastic, but they became embittered after discovering that the act would not completely exclude foreign fishing vessels and that the bill gave the federal government enormous power to manage the fishing industry. Many small fishers felt that the bill favored large corporations, and their private commentary was "vile, violent, and vituperative." Many favored no regulation at all; those who accepted regulation were widely divided on the kind; the closest they came to consensus was 97% opposition to a Federal quota system.(19)

Despite the uncertainties and dislike of the regulations, the fishers undertook major investments. The licenses issued for groundfishing in New England increased from 1200 in 1977 to 2191 in 1979. A majority of the fishers invested in larger vessels, in much more electronic equipment, and in additional kinds of fishing gear. A result was a large increase in versatility by a majority of the fishers who planned to exploit new grounds and to catch more species.

An evaluation of the reasons for the decisions indicated (1) that the fleet had been very old (many were over a normal design life of 20 years), (2) that the fishers had been doing well financially--in part because of a fortuitous short-term increase in abundance of a few stocks, (3) that the regulatory system of overall catch quotas meant that the largest catches went to the biggest and best-equipped vessels, and (4) that there were greater profits in groundfishing than in the other fishing. A few fishers took advantage of subsidized federal loans for their investments.[19]

But they continued their opposition to the federal management system which attempted to set annual quotas on each of the major species in the mixture of fish caught by the bottom trawl fishery. The species in the mixture varied independently in abundance with some requiring protection when others could be fished more heavily. The scientists advocated generalized solutions in three phases--multispecies management, optimum yield, and limited licensing--none of which was acceptable to the fishers in any practical way.[20]

After eight years the conservation purposes of the Act have been met only to a limited extent throughout the nation; most fisheries controlled under the Act have become fully or overdeveloped. A prediction has been made that much greater changes will happen in the near future. The tough decisions about allocation between commercial and recreational fishing have been largely avoided. The processing sector has not expanded significantly into the international market despite the large stocks available in the 200-mile zone. Nothing has been done to control overinvestment in the fishing (see Ref. 6-62).

9.4.2 Local Authority

The fundamental dilemma of finding the acceptable compromise between economic efficiency and equity in fishing comes to a focus at the local level. Here the problems are personal, the communications are directly among neighbors, the recent experience is vivid, and the solutions sought are immediate.

Local fishers frequently have strong beliefs in their exclusive rights to nearby stocks and fishing areas. They have long established formal or informal property rights to prime fishing space, found ways to exclude outsiders, and revealed private information on the resources only to colleagues.[21] A large portion of the ocean fisheries is community-based, yet the communities are seldom effectively involved in framing national laws or international agreements.

The problem is an old one. In Newfoundland, for example, a charter for the earliest settlement, adopted in 1610, prohibited the settlers from asserting jurisdiction over the fisheries; the English merchants were given precedence.(22) Nevertheless, the settlers under their first governor asserted jurisdiction; thus starting a dispute that has continued to the present (see Section 6.5).

Another example that has been well described is the Maine lobster fishery.(23) Here in the early 1970s the state laws required merely the licensing of fishers, protection of breeding female lobsters, and release of all catches except male or nonbreeding females with carapace length between 3 3/16 and 5 inches (81 and 127 mm).

The lobsters are caught in pots, and anyone with a license can legally set pots anywhere. But the established fishers have agreed among themselves upon fishing areas and numbers of pots in public waters, some of which are defended "to the yard." Anyone violating these understandings can expect to have buoy lines cut and thereby lose pots or to find a boat sunk. The fishers may also collaborate on fishing seasons, notably to cease fishing when other seasonal employment is available.

This political system is effectively allocating a scarce resource among the local fishers and, through the local understandings as well as the political strength of the fishers in the state legislature, achieves conservation of the resource. It is also believed by state fishery administrators that any attempt by the state to legally replace and enforce the detailed and complex local arrangements would result in prohibitive public administrative costs and would probably be ineffective.

The Maine lobster fishing is an example of adherence to perceived rights by a defined group of people who have made decisions about their own best interests in the use of the public resource. It is not open access; rather it is similar to rights of pasture, estover, and turbary in public areas and resources that are shared coequally. An example of another state attempt to curtail such rights is management of Raritan Bay, New Jersey, near New York City. Small-scale fishers have battled creation of large-scale commercial fishing for oysters and menhaden since early in the nineteenth century, and they continue to battle the predominance of recreational fishing in recent decades. They consider that their rights have been violated by the state and continue "piracy" to this day. They sustain their anger and their practice by the mythology of discrimination against them.(24)

The variety of problems spans too many academic disciplines for any narrow research program to be satisfactory. The biologist may investigate the extent and condition of the resource, the economist the opportunity costs, the legal scholar the freedom of access, the anthropologist the meaning of the fishing to the community, the political scientist the process of decision, and the historian the evolution of the fishing. How can any one be chosen?(25)

The inability of national governments to control the continuing increase in fishing effectiveness on stocks which are fished close to the maximum sustainable yield is probably the most widespread constraint on fishery development (see Section 4.4). In this situation the buildup of modern vessels first affects the livelihood of the traditional fishers using a stock (see Sections 6.3 and 6.5). As technical improvements continue, the fishers who cannot afford the new investments lag until only the most efficient vessels are profitable. More and more government help is required for social reasons.

9.4.3 International Relationships

The sovereignty of national governments over their fisheries must not only yield to the diverse problems of fishing communities but also to the uncontrollable habit of fish migrating across national boundaries. The fish of international rivers, lakes, estuaries, and practically all marine areas outside estuaries ignore the boundary formalities. The sovereign authority over an EEZ 200 miles wide rarely includes authority over the fish in it at all stages of life except by countries with very long coastlines.

The issues of international use of marine fisheries were summarized with great clarity shortly before a substantial consensus on international law was reached in Caracas in 1974.[26] Most of them apply today:

> It is clear that the international law of marine fisheries functions within a context of great biological, technological, economic, social and political diversity. Moreover, it is a context subject to rapid and often unpredictable change. This raises the question of which objectives the international law of marine fisheries can seek to realize. This question can be answered in general terms by stating, for example, that its goals are to maximize human welfare, to alleviate international conflicts, and to optimize world public order. However, a less abstract answer is desirable. Unfortunately, such an answer is very difficult to give in view of the extreme complexity of the problems...it is more realistic to take the view that the international law of marine fisheries does not have one single objective, but is confronted by a series of challenges. In this approach the need is to determine which challenges the international law of marine fisheries faces in specific areas, rather than what its single objective is.
>
> Six specific areas were identified: (1) the full utilization of the living resources of the high seas; (2) the conservation of these resources; (3) the economic efficiency of marine fisheries; (4) the allocation of the catch; (5) scientific research; and (6) interuse and intrause conflicts.

The quotation indicates the troublesome scope of national fishery problems that are commonly involved in fishery conventions, but, in addition, reaching agreement usually depends also on nonfishery issues. Countries always have diplomatic, trade, defense, or other political issues among them that influence the negotiating process for all or part of any one area. Nor should it be forgotten that wars and lesser violence have resulted from fishing disputes.

In summary, it seems clear that national market fishery policies tend to fall short of accommodating either the social and economic issues of the local fishing communities or the international fishery issues with neighboring countries, because neither fits well into the national competence. The national governments are always preoccupied with more important issues outside the fisheries and simply cannot become effectively involved in local details. The national governments must also deal with international fisheries in a mixture of issues few of which are related to local problems. Local organizations must become strong enough to deal with local issues within the context of national policy, and national organizations must be sensitive enough to delegate suitable authority to the localities. The experience with the rapidly changing fishery issues indicates that development of trusted institutions is a complex and long-term process.

NOTES AND REFERENCES

1. FAO, Inland Water Resources and Aquaculture Service, Fishery Resources and Environmental Division, A study of methodologies for forecasting aquaculture development. *FAO Fish. Tech. Pap.*, (248):1-47 (1984).
2. R. L. Schmied, The why and how of marine recreational fisheries development. *In* R. H. Stroud, ed., "World Angling Resources and Challenges: Proceedings of the First Angling Conference, Cap d'Agde, France, 12-18 September, 1984". Int. Game Fish. Assoc., Fort Lauderdale, Florida, pp. 355-364, 1985.
3. R. F. Hutton and J. H. Kutkuhn, Angling in the United States of America. *In* R. H. Stroud, ed., "World Angling Resources and Challenges: Proceedings of the First Angling Conference, Cap d'Agde, France, 12-18 September, 1984". Int. Game Fish. Assoc., Fort Lauderdale, Florida, pp. 83-100, 1985.
4. G. Pike, Birth of the Australian recreational fishing confederation. *In* R. H. Stroud, ed., "World Angling Resources and Challenges: Proceedings of the First Angling Conference, Cap d'Agde, France, 12-18 September, 1984". Int. Game Fish. Assoc., Fort Lauderdale, Florida, pp. 25-31, 1985.
5. V. G. Wespestad and S. M. Fried, Review of the biology and abundance trends of Pacific herring *(Clupea harengus pallasi)*. *In* W. S. Wooster, ed., "From Year to Year: Interannual Variability of the Environment

and Fisheries of the Gulf of Alaska and the Eastern Bering Sea. Washington Sea Grant Publ., Univ. Washington, Seattle, pp. 17-29, 1983.

6. A. D. MacCall, Variability of pelagic fish stocks off California. *In* G. D. Sharp and J. Csirke, eds., "Proceedings of the Expert Consultation to Examine Changes in Abundance and Species of Neritic Fish Resources, San José, Costa Rica, 18-29 April 1983." *FAO Fish. Rep.* (291, Vol. 2):101-112 (1983).
7. G. D. Sharp and J. Csirke, eds., "Proceedings of the Expert Consultation to Examine Changes in Abundance and Species of Neritic Fish Resources, San José, Costa Rica, 18-29 April 1983." *FAO Fish. Rep.* (291, Vol. 2), 555 pp. (1983). See numerous papers.
8. B. E. Brown, V. C. Anthony, E. D. Anderson, R. C. Hennemuth, and K. Sherman, The dynamics of pelagic fishery resources off the northeastern coast of the United States under conditions of extreme fishing perturbations. *In* G. D. Sharp and J. Csirke, eds., "Proceedings of the Expert Consultation to Examine Changes in Abundance and Species of Neritic Fish Resources, San José, Costa Rica, 18-29 April 1983." *FAO Fish. Rep.* (291, Vol. 2):465-506 (1983).
9. O. E. Sette, Biology of the Atlantic mackerel (*Scomber scombrus*) of North America: Part I, Early life history, including the growth, drift, and mortality of the egg and larval populations, *U.S. Fish Wildl. Serv. Fish Bull. 50*:149-237 (1943).
10. A. Bakun, Definitions of environmental variability affecting biological processes in large marine ecosystems. *In* K. Sherman and L. M. Alexander, "Variability and Management of Large Marine Ecosystems," Westview Press, Boulder, Colorado, pp. 89-108, 1986.
11. N. Daan, Results of recent time-series observations for monitoring trends in large marine ecosystems with a focus on the North Sea. *In* K. Sherman and L. M. Alexander, "Variability and Management of Large Marine Ecosystems," Westview Press, Boulder, Colorado, pp. 145-174, 1986.
12. G. Pontecorvo, Cost benefit of measuring resource variability in large marine ecosystems. *In* K. Sherman and L. M. Alexander, "Variability and Management of Large Marine Ecosystems," Westview Press, Boulder, Colorado, pp. 269-279, 1986.

13. V. M. Norton, M. W. Miller, and E. Kenney, Indexing the economic health of the U.S. fishing industry's harvesting sector. *NOAA Tech. Memo. NMFS-F/NEC-40*, Nat. Mar. Fish. Serv., Northeast Fisheries Center, Woods Hole, Massachusetts, 42 pp. (1985).
14. M. H. Glantz, Man, state, and fisheries: An inquiry into some societal constraints that affect fisheries management. *In* G. D. Sharp and J. Csirke, eds., "Proceedings of the Expert Consultation to Examine Changes in Abundance and Species of Neritic Fish Resources, San José, Costa Rica, 18-29 April 1983." *FAO Fish. Rep.* (291, Vol. 3):893-917 (1983). See p. 893.
15. *Ibid.*, p. 908.
16. *Ibid.*, p. 910.
17. Secretariat of the Expert Consultation on Strategies for Fisheries Development, Some considerations on strategies and policies for fisheries development presented in selected case studies. *In* "Case Studies and Working Papers Presented at the Expert Consultation on Strategies for Fisheries Development, Rome, 10-14 May 1983." *FAO Fish. Rep.* (295):16-37 (1983).
18. J. A. Gulland, Fishery management: New strategies for new conditions. *Trans. Amer. Fish. Soc.*, *107*(1):1-11 (1978).
19. J. M. Acheson, Government regulation and exploitative capacity: The case of the New England groundfishery. *Human Organization*, *43*(4):319-329 (1984).
20. S. Apollonio, New England fisheries management. *In* N. Krant, ed., "Science, Politics and Fishing: A Series of Lectures." Sea Grant College Program, Oregon State Univ., Corvallis, Oregon, pp. 57-67, 1981.
21. B. J. McCay, Systems ecology, people ecology, and the anthropology of fishing communities. *Human Ecol.*, *6*(4):397-422 (1978).
22. G. Stiles, The small maritime community and its resource management problems: A Newfoundland example. *In* D. M. Johnston, ed., "Marine Policy and the Coastal Community: The Impact of the Law of the Sea." St. Martins, New York, pp. 235-245, 1976.
23. J. M. Acheson, The lobster fiefs: Economic and ecological effects of territoriality in the Maine lobster industry. *Human Ecol.* *3*(3):183-207 (1975).

24. B. J. McCay, The pirates of piscary: Ethnohistory of illegal fishing in New Jersey. *Ethnohistory*, *31*(1):17-37 (1984).
25. D. K. Emmerson, Rethinking artisanal fisheries development: Western concepts, Asian experiences. *World Bank Staff Working Pap.* (423):97 (1980).
26. A. W. Koers, "International Regulation of Marine Fisheries".Fishing News (Books) Ltd., Surrey, England, 368 pp., 1973. Quote from pages 38-39.

Chapter 10

Overcoming the Constraints

The great hopes for fishery development raised by the extension of national jurisdiction over the EEZs during the 1970s have proved to be exceedingly difficult to realize. The results of the many national efforts may be likened to efforts by farmers to enclose and rear livestock by putting up notices about pasture borders, but without fencing the land or caring for the stock.

In addition to the problems of managing freely migrating stock pursued by vessels out of sight of land, many of the difficulties arise because of a widespread public perception of the value of improved technology, the effects of which are radically different in the three major kinds of fisheries. Fish farming, like agriculture, will respond to improved technology with increased production, employment, and economic growth. Recreational fishing can develop as part of tourism with allocation of a few fish stocks to it from the market fishing, where they can be much more valuable than if merely caught for the market. Technological improvements can be designed to increase the enjoyment of recreational advenures while *reducing* the catch if necessary to protect the resource. Market fishing, however, becomes trapped by technological improvements in fishing. These are necessary in open access fisheries for each fisher to maintain a share of the production, but they lead inevitably to overfishing, reduced employment, and reduced production in the longterm. And restriction of access to market fishing by national governments has proved to be exceedingly difficult. Nevertheless, development of market fishing is the major fishery development goal in most countries and is the concern of this concluding chapter.

Fishery development by governments is perhaps best described as the art of blending solutions to the biological, economic, legal, social, and political problems during the course of a continuing program. Each type of problem requires different expertise contributing to a different project over different times in a changing situation.

One must assume that if development were easy, the private sector would already have seized opportunities. On the contrary, development of a common property resource has been judged to be too difficult, or too risky, or requiring government involvement for political reasons.

10.1 PROBLEMS FOR ALL COASTAL COUNTRIES

The dilemmas of market fishing development already occur in most countries, developed as well as developing. They arise fundamentally from constant change in the technical, ecological, and social aspects of the fisheries that stem from the resources, the competition among fishers, the fishes' environment, the fish marketing, the competition with other protein foods, and the opportunities for alternative employment.

An example of the complexity emerged at a national seminar on fishery development held in India on April 9-11, 1982. A summary paper on the marine fisheries alone(1) noted the divergent opinions about the exploitable resources in the EEZ, the lack of data on commercial operations that prevented establishment of their economics, contradictory opinions about use of fiberglass reinforced plastic in canoes, the crashing foreign markets for tuna, the need for port development, the shortage of experienced fishers, the need for conservation measures and expanded enforcement patrols, the increasing coastal pollution, and the high cost of parastatal businesses. Similarly complex issues arose in the discussion of inland fisheries and fish farming.

All countries have both large- and small-scale fisheries. The small-scale fishers usually live in rural areas where opportunities for alternative employment are limited. There the lives of people are closely linked to the seasonal cycles of fishing. They usually follow family and village traditions that have existed for generations, and they continue to be involved as families in the related activities of catching, processing, and marketing. They are accustomed to the problems and dangers of living with the sea and with the expected as well as the unpredictable fluctuations in the resources. Consider, for example, that the fishers who live on the shores of the isolated fjords of Newfoundland probably have much in common with the fishers on the east coast of Thailand; the livelihoods of both were disrupted by the introduction of trawl fishing (see Sections 6.3 and 6.5).

Such people usually have a concept that the resources are as much theirs as the land used by farmers. And almost to a person, they object to their livelihoods being regulated by national governments for nationally determined objectives without their concurrence. They must endure the consequences of decisions in which they have had little input.

The overall difficulties of market fishing development become clearer by comparison with the primary objectives of agricultural development. These objectives, which will be elaborated with respect to fisheries in Section 10.3, have been stated on the basis of the World Bank experience.[2] They are (1) *growth* in output and productivity; (2) *sustainability* of production systems during the development of scientific farming; (3) *stability* of production and markets; (4) *equity* among those who own, earn, and market; and (5) *efficiency* in the use of land, labor, and equipment.

The pattern of successful fishery development efforts is to start with projects directed at urgent needs. Each project is a discrete unit of activity with clearly defined objectives which, when coordinated with related projects over a long term, contributes to the goals of a program.

Planning projects is an art and is beyond the scope of this book, involving as it does investments, services, institutional change, and policy change. Extended discussion of projects and their planning will be found in references.[3,4]

10.2 EXTERNAL ASSISTANCE

No worldwide summary is available for the domestically financed efforts to develop fisheries, but the amount of external aid is substantial, amounting to an average annual total from 1980 through 1983 of around US$400 million. About 95% of the total is described in detail in an FAO Fisheries Circular.[5] The balance, predominantly global, interregional, and regional projects, is not included.

The total of nearly US$400 million was an increase of about 50%, in constant dollars, since the early 1970s. The domestic assistance, which is known to be extensive in almost all of the developed countries, currently must be at least another US$1 billion.

The source of the external aid in 1981 was 52% from bilateral agreements with a donor country; 39% from multilateral arrangements with the World Bank, other United Nations agencies, and the Regional Development Banks; and 9% from the European Economic Community, the oil producing countries (OPEC), and technical cooperation among developing countries.

A large majority, 83%, of the assistance was for capital projects, especially for ports, vessels, and infrastructure. The balance was categorized as technical assistance, including research, advice, and training.

The assistance was distributed 47% for Asia, 25% for Africa, 15% for Latin America, 10% for the Near East, and about 2% each for the Caribbean and Oceania.

A major multilateral donor was the United Nations Development Programme, headquartered in New York. During the 1976-1980 period it supported 18 regional fishery projects, 15 small-scale fishery projects, and 30 fish farming projects, mostly through collaboration with FAO as the executing organization. These were categorized as resource surveys, small-scale fisheries projects, multipurpose development projects, and training projects. The total funding averaged about US$21 million annually during the period, about one-fourth of which were funds in trust from collaborating donors.(6)

The largest bilateral donor was Japan, which contributed 36% of the bilateral donations in 1981. Other countries which contributed from 7 to 11% each were Denmark, the Federal Republic of Germany, and Norway. Those which contributed about 4 or 5% each were Canada, France, Italy, The Netherlands, Spain, Sweden, the United Kingdom, and the United States.(5)

Japan increased its assistance greatly as developing countries were extending their jurisdiction, and, in many cases, the aid was part of a trade for fishing rights. About three-fourths of its aid in 1981 went to Asian countries. Its aid projects have been characteristically well organized, with training programs to teach the use of new equipment, and with spare parts and repair training for new machines. It has included several research and training vessels for various countries over the years.

Denmark, which contributed about 0.9% of its GDP to its total development assistance during the mid-1980s, supported agriculture, forestry, and fisheries with about one-third of its total contributions. It targeted small-scale fisheries and fish farming in order to increase supplies of desirable food and promoted wise use of the resources. It relied on its major competence in fishing, fish processing, fish farming, education, and consulting services to help more than 30 low-income countries but concentrated about two-thirds of its fishery efforts on Bangladesh, India, Kenya and Tanzania.(7)

The United Kingdom development assistance is entirely channelled through its Overseas Development Administration. The fishery matters are handled through the Ministry of Agriculture, Fisheries, and Food; and a number of academic, charitable, and voluntary institutions. Its aid "is aimed at the poorest people in the poorest countries, but giving due regard to the UK's own national interest where appropriate." Its fishery aid emphasizes manpower development but has included substantial investment in facilities.(8)

10.2.1 Strategies

Preparations for the FAO World Conference on Fisheries Management and Development in 1984 marshaled the long experience of FAO and its consultants in fishery assistance. It included two preparatory meetings by committees of experts who discussed strategies in depth.[9,10] The strategies were directed toward eight different goals, which were elaborated and endorsed at the final conference held between June 27 and July 6, 1984.[11] The principles and guidelines set forth represent the most comprehensive effort to that date to detail approaches to the problems of fishery development. They are listed, paraphrased, and greatly abbreviated for each of the eight goals in Appendix A.

These reveal the extraordinary complexity of fishery development. Many of the goals must be considered jointly and compromises made among them. Almost all of the principles and guidelines, when considered in the context of any country's government, involve multiple agencies that must be coordinated in long-term activities.

After drafting the strategies, the committee of experts at the April 1984 meeting drafted five programs for FAO action.[6] These were designed "to assist developing countries in further improving their individual and collective self-reliance in fisheries management and development." They were approved with minor changes at the conference.[11]

The action program on the *planning, management, and development of fisheries* stressed advisory services, regional collaboration, investment identification, and training. The program for *development of small-scale fisheries* stressed an integrated approach, investment and credit, and training and extension. The program on *aquaculture development* stressed regional research and training centers, linked to national centers, and a better information base. The program on *international trade* stressed regional and international market information services, a multilateral framework for consultation, and investment identification. The fifth program on the *role of fisheries in alleviating undernutrition* stressed improved utilization, research, and training. The total annual budget requested by FAO for execution of the programs was just under US$16 million in addition to all ongoing programs.

The FAO documents, from which this summary is taken, detail further the scope and complexity of international assistance projects in fisheries. The conferences, comprised as they were of representatives of sovereign nations, focused on national responsibilities but not on national policy development that would involve institutional changes within countries and require countries to manage continuing long-term programs in which each project builds on a predecessor.

10.2.2 Project Experience

The ultimate impact of international assistance programs for fishery development is largely unpublished. Each project has a limited duration, usually five years or less if not extended, and is focused on only a small part of any country's fishery problems. Each program requires a coordinated sequence of successful projects, and, because of the special government role in fisheries, effective government policies are vital.

Discovery and publication of poor project performance in any sector may cause political difficulties for either donor or recipient, so most assistance projects end with a perfunctory report of the work done and the money spent with little attention to their ongoing significance.

Such difficulties were apparent during the preparation of one of the most useful summaries of fishery project results.(12) The author reviewed six large projects, each specific to a country, but concealed the country identities. He details the lessons learned from each project experience which, if any summary is possible, amounted largely to the fundamental need for improvement in government administration of projects. The governments discovered the difficulties of project planning and administration, the operations that the private sector could do better, the rapid rate of change in the fisheries, the hazards of credit management, and the importance of improving national capability in administration and research.

One of the few agencies that is publicly critical of its own projects is the World Bank; it has discovered too many difficulties with its past fishery projects which have had Bank loans of about US$550 million.(13) The difficulties have included poor interagency coordination before investments, poor planning, too short a duration of technical assistance for small-scale projects, weak government administration, lack of experience in Bank appraisal missions, insufficient coordination among country agencies and donor agencies in sequential projects, and lack of appropriate credit structures for small-scale fishers.

A large majority of the Bank projects has supported new facilities such as vessels, ports, fish ponds and infrastructure, for which the return on investment has repeatedly been less than expected.(14) Investments in large-scale fisheries have frequently been inappropriate because of social conflicts. Investments in small-scale fisheries have been difficult to coordinate among the large numbers of fishers in diverse communities. Credit assistance has been especially difficult because fishers commonly receive frequent cash payments, may follow fishing to other ports, and, unlike farmers, have difficulty planning to make repayments at long intervals.

A major area where fishery resources have seemed likely to support large benefits to coastal countries is west Africa from Mauritania to the Congo. The sixteen countries were receiving about U.S.$18 million annually in 1983 from almost 20 donors. Some of the assistance programs had started 15 years before then. The result has been a disappointing rate of increase in catch of less than 1.5% annually by comparison with population growth rates of about 3%.(15)

The causes of the failure vary from place to place, but they have been related to overly ambitious projects, diffuse generalized objectives, lack of attention to political and social constraints, and ineffective project management. Projects were seldom planned as part of a continuing long-term program with provision for continuing local management and maintenance of sophisticated equipment. Some projects included establishment or enlargement of parastatal businesses, almost all of which have failed. Many projects were designed to help small-scale fisheries but failed because of price controls or an inability to compete with imported fish. Many donors, apparently with consideration of political gain, supported projects too large for the country's institutions. The authors conclude with the observations that far too little attention has been paid to evolution of policy and virtually no attention to evaluation of past failures.

The U.S. assistance to fisheries has emphasized delivery of scientific and technical advice and provision of training through its AID in the Department of State. Much of the support has involved U.S. universities, which have strengthened their own expertise and sent faculty to work in several dozen countries.(16)

An evaluation of AID projects pointed to one spectacular success in assistance to the Korean Fisheries from 1956 to 1974. This project focused on large-scale fishery restoration and development by an aggressive and experienced fishery industry which had been the third leading nation in fish catches prior to World War II. It included modernization of facilities and development of distant-water fisheries ending with Korea among the top 10 nations in fish production.

Unfortunately the ultimate impact of most of the rest of AID fishery projects is unknown. Final reports of projects detail what was delivered, but they provide little information on long-term changes created by the projects. A postevaluation system appears to be essential if projects are to contribute to long-term development.

10.2.3 The Project Cycle

Lending agencies may assist more effectively with long-term development programs than donor agencies when they maintain continuing contact with a succession of projects. The experience with a first project in a sector provides a basis of experience in the country for overcoming difficulties in the second; thereby starting a learning process for both the government and the lending agency.

The project cycle of the World Bank is exemplary.(3) A project is defined as a package of investments, policies, and institutional and other actions designed to achieve an objective at a given time. It will start with a plan that may include capital investment in facilities, provision of expert services, strengthening of institutions concerned with the project (including training of managers), and improvement of policies that may affect the project. It is, in effect, a disciplinary framework.

The cycle starts with *identification* by the country and the Bank of a high-priority problem that probably can be solved at costs commensurate with benefits. This is expected to result in definition of a specific objective. It continues with *preparation*, a refinement of the plan in all dimensions. Then, if it is a major project, a new type of project for the country, or otherwise not a continuation of work by experienced institutions, the project is *appraised* by an independent team. Next the project is *implemented*; it is financed, organized, managed, evaluated, probably modified, and completed. Last, and perhaps most important for a continuing program, the project is subject to *ex-postevaluation*.

The Bank describes itself as a *hands-on* lender. During the implementation stage, the Bank makes regular reviews of the project with emphasis on both project performance and fiscal management. Financial problems caused by inflation, fiscal policies, unanticipated technical problems, or foreign exchange can stall a project if not resolved quickly. Management and political problems are pervasive, and projects are commonly limited by the level of efficiency of the public administration. This shortcoming frequently limits or even negates the value of technical assistance to public institutions.

The ex-postevaluation provides the institutional learning experience for both the country and the Bank. The constraints, the successes, and the mistakes are identified. A comparison with similar projects is made. The findings contribute to improved performance in subsequent projects but, perhaps more importantly, also to confidence in the success of long-term programs.

10.2.4 Program Development

Effective programs must include a series of effective projects but must also be based on the expectation of rapid changes in the fisheries. Unfortunately, most investment decisions in vessels or shore plants are based on 10-20 year plans, but almost all fisheries experience substantial changes in fish stocks during each decade--frequently at intervals of only 2-5 years.

Country institutions can benefit not only from their own experiences but also from the experiences of their neighbors. The developing countries in any one region in particular, have similar institutional, technical, and social problems to overcome. Their learning experiences have been accelerated by participation in regional fishery development activities of the kind fostered by the UNDP/FAO program of Technical Cooperation among Developing Countries (TCDC).(6)

The South China Sea Fisheries Development and Coordination Program (SCSP) headquartered in Manila, started in 1972, was originally advocated by the Indo-Pacific Fisheries Council, supported by UNDP, and managed by FAO.(17) Subsequently it received substantial support from CIDA. The participating countries were Hong Kong, Indonesia, Kampuchea, Malaysia, the Philippines, Thailand, and Viet Nam. It has recently been terminated.

Its legacy, however, is increased collaboration among national, regional, and international fishery agencies based in part on an extensive series of technical papers concerned with fishery problems. One project was a survey of the nearby international pelagic resources, notably the tunas, which was followed by major increases in catch. The project spent about US$15 million provided by member nations, obtained additional foreign assistance of about US$35 million, and provided justification for international bank loans of about US$200 million.

Similar collaboration has continued to be fostered, however, in the southeast Asian area as well as in other parts of the world with assistance from ICLARM, headquartered in Manila. Organized in 1975 with funding from The Rockefeller Foundation, it began with the goal of providing long-term assistance in development of brackish water fish farming and artisanal fishing.(18) It started with promotion of collaboration among national, United Nations, and private organizations in use of all applicable biological and social sciences for fishery development.

The Rockefeller Foundation withdrew support in 1984, but the value of the continuing collaborative approach has been recognized by other donors, including development organizations from Australia, Canada, Denmark, New Zealand, the United Kingdom, the United States, the Federal Republic of Germany, as well as United Nations agencies, The Ford Foundation, and the Skaggs Foundation. Work is continuing with emphasis on the Southeast Asian and South Pacific areas.[19]

A regional learning program for west African countries has been promoted by the Committee for the Eastern Central Atlantic Fisheries (CECAF). This was established by FAO in 1967 but did rather little until UNDP financing was implemented in 1975. Membership now includes 21 African nations from Morocco to Zaire and a number of non-African nations that fish the same area.[20]

The immediate objectives were to improve fishery statistics, develop a system for resource monitoring and evaluation, enhance country abilities to manage fisheries, stimulate regional and subregional collaboration, and train personnel.

The contribution to the statistical and resource evaluation work has been the major contribution of the program. This has provided an essential basis for development which has not been realized because of sociopolitical difficulties in most of the countries.[15]

Another regional project has focused on development of small-scale fisheries. The Bay of Bengal Program (BOBP) emerged in 1974 as an initiative of the Swedish International Development Authority (SIDA). An overall Indian Ocean Programme had been conducted for several years by FAO with financing from UNDP, during which a broad background of information had been collected about the entire Indian Ocean fisheries. The need for the BOBP was identified as provision of institutional support services in the fields of technology development, education, and training.[21] The program, under FAO direction, began operations in 1979 from headquarters in Madras, India, with the country membership including

Bangladesh, India, Malaysia, Sri Lanka, and Thailand. The mode of operation is to support the national staffs.

In 1984, the program emphasized marine fish farming, especially shrimp at several different locations, and noted the striking success of cage culture of seabass and grouper in Thailand, where the practice had been demonstrated earlier and had stimulated extensive private investment.(22) Other work included tests of cottage-level net-making systems, improved set nets, improved trawl nets, and sail-powered fishing craft capable of landing on beaches. A major activity was extension service, especially to women and fisher's families. The report noted the rapid extension of credit to fishers that had followed the special training of extension officers who had developed communication between fishers and bank officials. Work had also continued on informal education for children in fishing villages. Numerous projects had failed, especially in the shrimp farming, but the reasons were understood and provided a basis for planning other projects.

Other regional programs with strong participation from countries with common problems are also gradually having an impact on institutional performance. Another example is the Forum Fishery Agency, comprised of southwest Pacific countries, with vigorous financial support from Australia and New Zealand. The key to the success of all of them is apparently to bring the people from national institutions together at intervals over a long term so they can compare learning experiences.

A similar system of regional centers for development of fish farming has been organized by UNDP/FAO as the Aquaculture Development and Coordination Programme (ADCP). Seven centers were established between 1978 and 1982 in Nigeria, Brazil, India, Thailand, the Philippines, China, and Hungary.(23) The coordination of the program is financed by UNDP; the operating expenses of each center by the host country.

The major strategic elements of the program have consisted of (1) identification of farming systems that have potential for widespread adoption in each region, (2) wider application of known technologies, and (3) the support services of adaptive research, training, and information.

The formation of these centers was due to recognition of the primary importance of technical information in the development of fish farming. Training of senior aquaculturists is promoted regionally; in many cases through collaboration with universities. Training of technician/extension workers is promoted nationally, with emphasis on the local farming systems and in the local languages. The centers also support distribution of information obtained from international bibliographic services as well as local information of immediate practical importance. In addition, the program has provided special assistance for planning development projects in both small- and large-scale fish farming, in transitional assistance for small-scale fishers, and in furnishing critical supplies such as seed and feed.

National fish farm extension services have been expanding rapidly. By 1982, information was available on the extension programs of 32 countries.(24) Some were still in the planning stage but most already supported training and research components. Several had also received additional bilateral assistance, especially from the U.S. Peace Corps.

10.3 THE POLICY DILEMMAS

Development of national market fishing policies in most countries faces some almost irreconcilable issues that are inherent in sustaining use of the commons, and in balancing international, national, and local interests. Since many people think of production of food from market fishing as similar to farming, a comparison with the government objectives of agricultural development (section 10.1) will emphasize the differences and complexities.

An overall objective of *growth* in output and productivity means to produce more food as well as to create jobs, to increase incomes, and to generate more revenue to invest in agriculture and other sectors. Here, in this overall objective, the differences in national and local interests arise. Conservation of fish stocks is almost universally a national or intermediate level government function, whereas growth in economic benefits of market fishing depends on local initiatives. Farming, in almost all countries, is much more important and the production systems are entirely within national boundaries, whereas a majority of market fish stocks must be shared among countries. Furthermore, market fishing is almost always restricted to small regions of each country, whereas farming is commonly widespread, albeit with different crops.

The second objective is *sustainability* of the production systems during the growth. This requires conservation of land and water for farming, and fish stocks and water for market fishing. Conservation programs for land and water tend to be local or regional, those for market fish stocks tend to be national or international. Furthermore, farm production is usually expected to be sustained by continuation of past practices and grow with continuing innovation, investment of profits, and upgrading of technology. Most market fishing, however, cannot even be sustained economically by continuation of past practices.

A third objective is *stability*. Both farming and market fishing are subject to serious weather and ecological changes as well as the trade cycles. But farmers control the reproduction of both plants and animals, whereas market fishers depend entirely on natural reproduction which, for many major fisheries, is extremely variable and unpredictable.

A fourth objective is *equity*. In farming this is often determined by who owns and earns what at the outset and by the functioning of the marketplace. The practices in most countries have been accepted for generations. In market fishing, however, it is judged both by access to the resources and the marketplace. The ease with which production per person from fishing can be multiplied by technical innovations leads quickly to exceeding the productivity of the stock, to ruining livelihoods of the traditional fishers, or to both.

In farming (or in industry) the introduction of technology to increase production per person would be promoted as progress toward the objective of *efficiency*. But in market fishing, where the fishers lack ownership of a quota or license, technical progress leads inevitably to overinvestment, to overfishing with damage to the resource, or to both. Worse, failure to reduce costs in market fishing leads to inability to compete with imported fish, to compete in international markets, or to compete with other protein foods.

Because of these problems, some fishery developers have advocated promotion only of appropriate technology in the lesser developed countries. In so doing, they joined a movement that had begun in the 1960s after technology transfer to numerous countries had been found to be too sophisticated for their use. Unfortunately, the technology of fishing, processing, and marketing has been changing rapidly, and maintenance of economic efficiency requires continuing technical change as well as coordinated social and political changes. The appropriate technology movement appears to have been too simplistic.[25]

10.3.1 Promising Directions

Any socioeconomic rationalization of excessive market fishing requires abrogation of the widespread right of piscary in the commons. The first move in this direction was taken long ago by governments that permitted fishing and marketing only by those qualified and licensed for reasons of nationality or place of residence. Use of this traditional system provides an accounting of the fishers, and, since it preserves the right of piscary to all within each government's jurisdiction, it has been widely accepted as equitable. The licensing systems are fundamental to management for conservation, which has introduced widespread inefficiencies through restrictions on seasons, size of fish, kind of gear, and locality of fishing. The conservation restraints have always had to be framed around elements of equity and have occasionally simply prohibited efficient gear operated by large businesses.

A move in the direction of combining conservation (i.e., sustainability) and economic efficiency has been to limit the number of licences. This has been advocated on a theoretical basis by economists for many years, but public decisions to do so have generally lacked the confidence of the fishers who have foreseen many problems of equity with little evidence of practical success.

Experience in several programs (mostly in Canada and the United States) was reviewed at a conference on limited entry in 1978.[26] Most of these did not clearly separate the goals of conservation and efficiency, nor did they deal in depth with the complex goal of equity. Some dealt with other social goals, such as redistributing income from corporations to individuals, increasing the bargaining power of fishers, and individual ownership of vessels. Collectively, they revealed a range of complexities that can be considered steps in a process of limiting entry.

The first and simplest step is merely to limit the number of licenses, either to the number of fishing units already participating in a fishery for a designated stock or to a smaller number achieved by government buy-back of units. The second step is to assign a property right in the licenses, which could then be transferred for a price. The third step is to add to each license a property right in an individual transferable quota (ITQ) or share of the estimated sustainable catch that could be taken under a conservation program.

These steps have been supported by biological and economic theory that presents market fishing as a relatively simple and coherent activity. It is not--fishers usually operate individually and treasure their independence. Each one expects fishing to contribute to a livelihood in a personal way. Many are part-time, fishing during a favorable season to supplement other income. Even full-time fishers almost always shift from one kind of fishing to another during the seasonal rhythm of fish runs and during slumps in a favorite stock. In fact, the most successful are commonly those with the ability to shift quickly and efficiently from one stock to another.

Consequently it was quickly discovered that each of these steps created controversial issues of equity depending on the circumstances of each fisher and each stock at the time of decision. Practically all fishers recognize conservation as vital for their long-term fishing, but the conservation issues tended to be mixed with issues of who could fish and who could not. Even limiting the licenses to those fishing on a designated stock created issues of discrimination. Some fishers depended exclusively on that fishing, others shifted regularly to other stocks. Some fishers, without ownership in a license, wanted to stop fishing but had nothing to transfer to another member of the family, and no benefit from relinquishing the license. A conservation restriction on one kind of gear or an area fished only by part of the fishers raised further issues of equity.

Most of these issues continued when licenses were reduced from the previous level through government buy-back programs. Although some fishers had an opportunity to sell their vessels, the fishers with diversified equipment had difficult choices. Furthermore, most of the buy-back programs resulted merely in the vessels being moved to other political jurisdictions.

If the licenses could be transferred to another vessel under the same owner, merely limiting the number of licenses allowed fishers to turn to larger vessels and better gear--the fishing effort continued to increase. Even limits on vessel size, horsepower, and gear size have not stopped a continuing increase in effectiveness.

The difficulty of controlling the fishing effort by restricting licenses is revealed by an example of a major increase in effectiveness, which occurred merely through a change of hooks used by the northeast Pacific halibut fishery (in which the Canadian fleet had limited entry, the U.S. fleet did not). The traditional hook had been shaped like a letter J, but trials of a circle hook in the late 1970s were so successful that it had been widely adopted by 1984. Comparative studies revealed that the catch-per-unit-of-effort by the circle hooks was 2.2 times greater than with the J hooks.[27] Such a change in gear did not require a large investment, but it essentially halved the time that the fishers needed to fish in order to catch their overall quota.

The second step in programs to limit fishing effort was a major one of conferring a property right in a license that could be bought and sold. Such programs were introduced into the salmon fisheries of British Columbia and Alaska during the 1970s.

The program in British Columbia revealed the complexity and unanticipated difficulties of the action.[28]

Special licenses were issued to vessels in 1969 based on those qualified by landings in 1967 or 1968. These were renewable annually, based on participation in the fishery at least every other year. The initial cost of the licenses was nominal. In a second phase starting the next year, the license fees were substantially increased in order to finance a buy-back program.

These fisheries, which include large efficient purse seiners, smaller gill netters, and trollers had been in an advanced stage of obsolescence and overcapitalization in the late 1960s. The objectives were to increase levels of income to the average national wage, reduce the overcapacity in the fleet, and improve the mangement of the resource. The buy-back program removed a small percentage of the less-efficient vessels.

By the late 1970s it was clear that none of the objectives had been achieved. Improvements in the remaining vessels and their operations had increased the total fishing capacity. The ownership of the fleet had been transferred to wealthier fishers. The competitive relationship among gear types in catching and marketing had been changed. The management of the resource had not been improved.

Alaska's experience was more successful, after its program for salmon and herring fisheries was introduced by a legislative act in 1973.[29] Licenses (permits) were issued to individuals (commonly captains who fished with family members) using specific types of gear in each of 29 different fisheries. The licenses were freely transferable, but no person was allowed to own more than one for the same type of gear in each fishery in each area.

The concept of limited entry was highly controversial from the time of its proposal, but it was supported by a large proportion of the fishers who sensed a disaster ahead in the major natural resource of the state prior to the development of petroleum. It was discussed knowledgeably by a state legislature with a significant number of members familiar with the fisheries, and introduced to the public through an extensive series of public hearings prior to enactment.

It withstood a repeal referendum on the ballot in 1976. But by the late 1970s, several hundred fishers, originally excluded from licenses, had appealed and several dozen court cases were pending. Many concerns remained in the early 1980s,[30] including safeguarding the livelihoods of native Alaskans, and reducing speculative activity, while preserving opportunities for entry into fishing after license values had escalated--in a few instances to more than $100,000.

The programs in both British Columbia and Alaska have no doubt become more secure because of escalation of both salmon production and salmon prices. By the mid-1980s the total production was close to all-time records and the stocks were being carefully conserved. Salmon prices were also at all-time highs, due primarily to rapid increases in foreign markets.

Another country with favorable experience in limited entry is Australia.[31] Its rock lobster fishery in Western Australia developed rapidly in the 1950s, and evidently approached maximum sustainable yield about 1960. The fishers, who had become alarmed by a major drop in catch-per-pot, readily supported the concept of limited entry. The number of vessel licenses was fixed at about the number actually in operation, and the number of pots was fixed according to boat length. Nevertheless, the effectiveness of the operations continued to increase. By the late 1970s the annual exploitation rate was estimated as about 70% of all legal-sized lobsters in the stock. The fishing licenses became valued at more than A$20,000.

A very different experience occurred in the Western Australian shrimp fisheries.[31] Shrimp stocks were known to exist in three remote areas that had been little fished. Exclusive rights were offered in 1963 to two shorebased processing companies in each of the three bays. The number of fishing vessels was restricted in each of the bays according to their size. The objectives were to insure the economic viability of the fishing units, of the processing units, and conservation of the resources. The ensuing operations have been very stable and apparently very profitable--approaching the economic ideal of private ownership. Few sales of fishing licenses had been recorded by 1977, but fishers then valued them at between A$150,000 and A$200,000.

An important consideration in evaluation of the success of both of these limited entry programs is the lack of diversification by the fishers. Apparently none had significant earnings in any other kind of fishing. Furthermore, government action was taken during a window of opportunity when the limitation was readily accepted by the participants. And economic growth of the fisheries was substantially assisted by export of a majority of the catch at rapidly increasing prices. Had the prices not increased, they probably would have been in economic difficulty.

These early experiences have been followed by limited entry programs in other fisheries, which have been readily accepted on the basis of favorable public impressions gained from the economic success and improved government relationships with industry. The latter has included the six states, which have responsibility within three miles, as well as the Commonwealth of Australia, which has responsibility outside that zone. Generally the programs have been adopted after collection of good information on the resources and the fishing, after careful coordination of state and commonwealth programs, and after extensive dialog with the people affected. Economic objectives are now considered along with conservation and socio-political objectives in the management of all market fishing.(32)

Most of the current efforts in the Australian program are going toward the third step in rationalization, that of assigning individual quantitative rights. This is being done with consideration of government financial assistance during transitional periods and with the announced intention of eventually assessing the industry for up to half of the costs of managing the fishing.(32)

Despite the encouraging progress, Australia has far to go in rationalization of its market fishing. It has an EEZ and continental shelf outside the EEZ with a total area of about 1.6 times its land area. It lacks a maritime tradition of using its surrounding seas--having focused on its inland resources.(33) Several major fisheries under its jurisdiction have been developed under license to foreign fishers. It has also imported a steadily increasing proportion of the fish consumed--almost 60% in 1983-1984. Part of this was fish products not available from Australian stocks, but part was also fish products produced elsewhere at lower cost. The overall value of the production to Australian fishers has recently exceeded A$500 million annually, of which about A$400 million is exported--mostly prawns and lobsters. Australia also has a major marine recreational fishery on which fishers spent an estimated A$2,200 million in 1983-1984.(28)

The overall application of quantitative rights to fisheries has been examined theoretically and by comparison to other rights in the commons, such as grazing on public lands, water rights, and pollution discharge rights.(34, 35) Many complexities are recognized. Should a right be to a specific quantity of fish or to a specific share of a fluctuating catch? Should it pertain only to one stock, only to one kind of gear, to mixed stocks as in trawl fisheries, or to different gears used on different stocks? Should a right be annual or long-term? Should it be given to all who happen to be fishing at the start, or auctioned to high bidders? Should it be transferable as a unit or in any fraction? Should it be transferable only to family members or partners, or only to other individual fishers, or to large businesses that could buy multiple rights? Can the system be enforced to keep catches within rights? If the fishers are in economic difficulty before the system is adopted, how should governments help financially at the outset?

The purpose of transferable rights is to let fishers make decisions through a normal market mechanism as with other property, but at the outset each fisher will face such complexities without experience and almost surely with suspicion of government effectiveness.

Quite another approach has become singularly successful in Japan after a long period of public involvement. In the early twentieth century, the Japanese fishers of Hokkaido lived poorly in remote villages where they were at the mercy of usurious markets and creditors. One man, Takatoshi Ando, began a campaign in the 1930s to strengthen the fishing cooperatives. He found an existing cooperative, persuaded local merchants to ease the debt burden of its fishers, and persuaded the cooperative to market the catch of its fishers. In its first year, the cooperative was able to sell its fish for prices about one-third more than the prices received in other fishing communities.[36]

Mr. Ando next went to the largest fishing company in Japan, Nippon Suisan (recently on the Fortune 500 list of the largest corporations in the world outside the United States), to obtain a larger loan. This connection and effective management of the cooperative business, established more confidence in fishing cooperatives, which was reinforced by strong national support of cooperatives after the War by the Supreme Commander of Allied Powers during the occupation.

With this political strength, Mr. Ando broadened the function of the cooperatives to focus on women's activities, family life, savings plans, housing, and community improvement. He took advantage of all prefectural and national movements to strengthen cooperatives and established a process of continuing development.

Now the fishing (and fish farming) cooperatives of Japan are organized into local associations, prefectural associations, and national associations which collectively manage an immense business and market at wholesale level about half the fish sold in Japan. They have great political strength at prefectural and national levels. Their membership includes a large proportion of the fishers and fish farmers who must belong to enjoy the fishing rights granted only to cooperatives in the inshore fisheries. Rights in the offshore and distant water fisheries are granted partly to cooperatives and partly to other type businesses. Entry to all fisheries is limited and occasionally has been reduced. This means cooperative management of virtually all of the coastal fisheries and a large but minor part of the offshore and distant water fishing (see Section 6.6).

This system, which has been developing vigorously for about half a century, now faces the political complexities of reducing entry into more of the fishing, of encouraging control of pollution and improvement of the marine environment, and of working closely with the government to modify the fishery law with respect to fishing rights and licenses.

10.4 POLICY CHANGE

Although the biological and economic theories of how to manage fisheries have been extensively refined, they are inadequate to explain many issues raised by fishers who will be affected by the transition to ownership of a right to use a resource of the commons. It is clear that policies must change, but we have little theory or experience to guide us on how to make changes. We continue with familiar policies that include customs, laws, and government institutions, despite widespread recognition that they are unsatisfactory.

Policymaking in modern market fishing must be a blend of science and politics. The sciences, especially biological, have been widely accepted as a basis for conservation decisions, but scientific contributions will probably have much less impact than politics on new policies for development decisions.

Perhaps the best approach to understanding the process of policy change in the fishing is to examine the prevailing conflict between analysis (i.e., science) and politics in all policymaking, as discussed by a political scientist.[37] It is an eternal struggle between facts and value judgments, between analysis and political interaction.

Fishery biologists have had major influence on policy, because conservation goals are widely accepted as is the necessity for scientific inputs to conservation decisions. Yet the new emphasis on growth, equity, and economic efficiency has had relatively little support from the sciences for what are much more complex goals than the conservation goal. These goals involve conflicts of values that change from fisher to fisher, village to village, region to region, and nation to nation. These are understood in a general way by analysts, but not in the particulars as seen by the fishers. The workable approach appears to require development of local institutions through a continuous process of incremental change.

The difficulties appear regularly in all fishing countries, developed and less developed, with remarkable similarity. The key to their solution in the long term lies in design of new problem-solving institutions, so perhaps a start should include examination of the broad principles of the relationship between government and people in the development process.

One such examination directed especially at general development problems suggests that development is essentially a modernization of societal structure so people have more influence over policies that affect their lives. This requires understanding of how best to make decisions at different government levels and among trade organizations, cooperatives, unions, professional associations, and others.[38] It suggests four principles.

The first principle is to identify and define the functions that should be performed at different levels of government from national to local. In the fisheries this must emphasize the complexities and the trends, with special attention to functions better handled at local levels. It should include the research system with its stock evaluation proceedures, the regulatory processes, the enforcement mechanisms, the investment decisions in infrastructure and all aspects of the fisheries, the environmental problems, and the government assistance programs. Special attention must be given to new development functions that will involve issues of equity and economic efficiency.

The second principle is that national governments must help define the policy framework at each level with advice and assistance as needed. This must include delegation of authority to regional and local institutions with attention to necessary linkages among levels. It must also include delegation of authority to finance projects, institutions, and infrastructure. In the fisheries, this must include identification of legal and institutional constraints, better consultation among levels of government, and perhaps temporary financing of local institutions by national governments. It will probably also include new laws, new institutions, and new organizations that will be required for development.

After functions, authority, and institutional structure have been agreed upon, the third principle is for the national government to set minimum standards of performance and enforce them with a system of audit, review, or whatever is necessary. In the fisheries this would include coordination among jurisdictions (national as well as local) sharing the same stocks, the statistical systems for recording catches, and the enforcement systems covering all regulations and agreements.

The fourth principle is to insure the continuity of the system with assistance to leaders of new institutions and with support of planning activities. The leaders will be the people who need to develop loyalties that will make a local problem-solving system work. They are vital to any organization and may require training, periodic assistance with organizational problems, change in laws or regulations, and temporary financing. The planning, in these days of worldwide marketing, may require international intelligence systems and close cooperation of national and local institutions.

These general principles are consistent with more pragmatic studies of rural development in less developed countries.(39) A prevailing problem in project management arises from a project, under control of a single institution, which requires skillful influence on and appreciation of all related institutions that may affect project performance. It is suggested that the organizational design should be concerned with three levels of institutional environment; (1) the organization's own internal "controlled" environment, (2) an external "influenceable" environment--the other organizations that provide its inputs or receive its outputs, and (3) the organizations that cannot be controlled but which must be "appreciated."

The effectiveness of the organizations depends critically on the public service employees. Although the developing countries have been expanding the number of their employees very rapidly and average about 80% as many per person in their central governments as the developed countries, their intermediate and local levels have only about 10% as many per person.(40) Their local governments also tend to be fragmented, poorly organized, and either invisible or largely ceremonial.(41) In addition, the public service at all levels in the developing countries has major shortages of professional and technical skills, few relevant training programs, and largely inadequate personnel policies.

Cooperatives, which are similar to local governments in many ways, have similar organizational problems. The relative success of fishery cooperatives has been spotted with many failures. A recent review of them[42] stresses the importance of leadership and communication in order to maintain internal support, overcome external sabotage by business competitors, and solve technical and financial problems. Their relationship to government is critical, because the fishery resources must be conserved and the right of piscary sooner or later must be limited. Both the cooperative and its government must be committed to long-term evolutionary development.

10.5 PARTING OBSERVATIONS

Market fishing development is in trouble in almost all countries, developed and developing, and at all levels from artisanal to the most sophisticated. It is obvious that biological, economic, and technological fixes are not working. The government institutions needed to foster sustainable, equitable, and stable growth, while enhancing economic efficiency in use of a resource of the commons, are lacking almost everywhere.

Clearly, such development must be approached as a long-term process of social evolution, rather than merely programs of conservation supported by fishery science and directed by fishery scientists. It will be an extraordinarily dynamic process, varying according to changes in the resources, changes in the role of fishing in the lives of people catching and using the fish, changes in the ability of processors and marketers to compete, and changes in the ability of governments to conserve the resources of the commons. The days are over in which it depended on the traditional ability of fishing businesses to discover, harvest, process, and market new resources as familiar ones became depleted.

Let us emphasize the priority of the social problems by noting that the total catch of the world's fishing for conventional species of the commons appears to be nearing a limit of about 100 million tons annually and that harvest of unconventional species such as krill will almost surely involve high technology by a few very efficient operations. Let us consider also that modern fishing technology can easily equip fishers to catch more than 100 tons each per year from any waters capable of producing that much.

Thus a completely technological approach to development of the fishing in order to produce food from the waters would eventually require less than 1 million fishers. This would then be at the expense of employment and family food for perhaps 30 million existing artisanal or small-scale fishers, and would foreclose recreational use of the fish by perhaps 200 million fishers.

The prevailing long-term development issue is how to establish trusted institutions that can allocate and manage the right to fish with acceptable social benefits.

NOTES AND REFERENCES

1. M. D. Reddy, Highlights of the discussion on EEZ and new opportunities. *In* "Fisheries Development in India: Some Aspects of Policy Management." Concept Pub. Co., New Delhi, pp. 529-537, 1983.
2. W. C. Baum and S. M. Tolbert, "Investing in Development: Lessons of World Bank Experience." Oxford Univ. Press, New York, 610 pp., 1985. See p. 88.
3. *Ibid.*, See Chapter 2 and Part III.
4. R. M. Lawson, "Economics of Fisheries Development." Prager Publ., New York, 283 pp. 1984. See Chapter 6.
5. H. Josupeit, A survey of external assistance to the fisheries sector in developing countries, 1978 to 1981 (excluding global, inter-regional and regional projects). *FAO Fish. Circ.* (755), 40 pp. (1983). Augmented by preliminary data for 1982 and 1983 provided by J.-L. Gaudet of FAO.
6. UNDP, "Fisheries Development, Review of Support by the United Nations Development Programme." UNDP, New York, 48 pp. + 9 tables and 18 appendices, April, 1986.

7. J. G. Jensen, "Danish Development Assistance in the Field of Fisheries." Prepared for the Fish. Dev. Donor Conf., Paris, 13-15 October, 1986. 21 pp., 1986.
8. Overseas Development Administration, "UK Fisheries Assistance Overseas: A Brief Accound of UK Policy, Programmes and Capacity for Assistance in the Fisheries Sector." ODA, 25 pp., August, 1986.
9. FAO, Report of the expert consultation on strategies for fisheries development (with particular reference to small-scale fisheries), Rome, 10-14 May 1983. *FAO Fish. Rep.* (295):1-35 (1983).
10. FAO, Draft strategy for fisheries management and development and associated programmes of action. FAO World Conference on Fisheries Management and Development, Rome 27 June to 6 July 1984. *WCFI/84/3*. 37 pp. (April 1984).
11. FAO, "Report of the FAO World Conference on Fisheries Management and Development, Rome, 27 June - 6 July 1984." FAO, Rome, 60 pp. + appendices, 1984.
12. W. H. L. Allsopp, "Fishery Development Experiences." Fishing News Books Ltd., Farmham, Surrey, England, 158 pp., 1985.
13. E. Loayza. The World Bank and fisheries development. *Infofish Marketing Digest*. *4/84*:24-28 (1984).
14. A. Sfeir-Younis and G. Donaldson. "Fishery Sector Policy Paper." World Bank, Washington, D.C., 79 pp., 1982.
15. M. A. Robinson and R. Lawson, Some reflections on aid to fisheries in West Africa. *Mar. Pol.*, *10*(2): 101-110 (1986).
16. Marine Technical Assistance Group, "An Evaluation of Fishery and Aquaculture Programs of the Agency for International Development." U.S. Nat. Acad. Press, Washington, D.C., 161 pp., 1982.
17. Allsopp, *op. cit.*, Chapter 8, pp. 108-121.
18. ICLARM, "Program Development Statement." International Center for Living Aquatic Resources Management, Manila, Philippines, 35 pp., September, 1976.
19. J. L. Maclean, L. B. Dizon, and M. S. M. Sadorra, eds., "ICLARM Report, 1985," International Center for Living Aquatic Resources Managment, Manila, Philippines, 117 pp., 1986.

20. Allsopp, *op. cit.*, Chapter 9.
21. FAO, Project for the development of small-scale fisheries in the Bay of Bengal--preparatory phase. *Indian Ocean Prog, Dev. Rep. 44*(1):1-27 pp., (1978).
22. BOBP, Report of the 9th meeting of the advisory committee, February 25-26, 1985, Bangkok, Thailand. *Bay of Bengal Programme, Development of Small-Scale Fisheries, Rep.* (22):1-41 (1985).
23. Advisory Committee on Marine Resources Research (ACMRR), Report of the second session of the Working Party On Aquaculture, Rome, 21-25 March 1983. *FAO Fish. Rep.*, (287):1-14. (1983)
24. M. S. Chakroff, Aquaculture extension services review. *FAO Fish. Circ.*, (747):1-25 (1982)
25. R. S. Eckaus, Appropriate Technology: The Movement Has Only a Few Clothes On. *Issues in Sc. and Tech. III*(2):62-71 (1987).
26. E. B. Joseph, Experience with limited entry: A review of contributed papers. *In* R. B. Rettig and J. J. C. Ginter, eds., "Limited Entry as a Fishery Management Tool," Sea Grant Publ., Univ. of Washington Press, Seattle, pp. 13-21, 1980.
27. Int. Pacific Halibut Commission, Population assessment, *In* Annual Rep., 1984, pp.21-25 (1985).
28. G. A. Fraser, License limitation in the British Columbia salmon fishery. *In* R. B. Rettig and J. J. C. Ginter, eds., "Limited Entry as a Fishery Management Tool," Sea Grant Publ., Univ. of Washington Press, Seattle, Washington, pp. 358-381, 1980.
29. A. Adasiak, The Alaskan experience with limited entry. *In* R. B. Rettig and J. J. C. Ginter, eds., "Limited Entry as a Fishery Management Tool," Sea Grant Publ., Univ. of Washington Press, Seattle, Washington, pp. 271-299, 1980.
30. T. A. Morehouse, and G. W. Rogers, "Limited Entry in the Alaska and British Columbia Salmon Fisheries." Inst. of Social and Economic Res., Univ. of Alaska, Fairbanks, Alaska, 291 pp., 1980.
31. T. F. Meany, Restricted entry in Australian fisheries. *In* R. B. Rettig and J. J. C. Ginter, eds., "Limited Entry as a Fishery Management Tool," Sea Grant Publ., Univ. of Washington Press, Seattle, Washington, pp. 391-415, 1980.

32. B. Lilburn, "Management of Australian Fisheries: Broad Developments and Alternative Strategies." Internal report, Australian Fisheries Service, Dept. of Primary Industry, Canberra, 13 pp., April 1986.
33. A. Bergin, Australian ocean policy--the need for review.*Mar. Pol.*, *10*(2):155-158 (1986).
34. D. G. Moloney and P. H. Pearse, Quantitative rights as an instrument for regulating commercial fisheries. *J. Fish. Res. Board Can.*, *36*:859-866 (1979).
35. P. H. Pearse, Fishing rights, regulations and revenues. *Mar. Pol.*, *5*(2):135-146 (1981).
36. R. Hamlisch and M. Hotta, What happened in Hokkaido? In "Symposium on the Development and Management of Small-Scale Fisheries: *Proceedings of the Indo-Pacific Fisheries Commission*, 19th Session, Kyoto, Japan, 21-30 Man 1980" FAO, pp. 395-401, 1980.
37. C. E. Lindblom, "The Policy-Making Process," 2nd ed. Prentice Hall, Englewood Cliffs, New Jersey, 131 pp., 1980.
38. E. Owens and R. Shaw, "Development Reconsidered: Bridging the Gap Between Government and People." Lexington Books, Heath and Co., Lexington, Massachusetts, 190 pp., 1974. See Chapter 2.
39. W. E. Smith, F. J. Lethem, and B. A. Thoolen, The design of organizations for rural development projects--A progress report. *World Bank Staff Working Pap.*, (375):1-48 (1980).
40. S. Ozgediz, Managing the public service in developing countries. *World Bank Staff Working Pap.*, (583):1-84 (1983)
41. G. Cochrane, Policies for strengthening local government in developing countries. *World Bank Staff Working Pap.*, (582):1-57 (1983)
42. Plunkett Foundation, Small-scale fisheries cooperatives--some lessons for the future. Committee for the Promotion of Aid to Cooperatives, FAO, Rome, *COPAC Occasional Pap.*, (2):1-51 (1984).

Appendix A

Summary of Goals, Principles, and Guidelines for Fishery Development

The FAO World Conference on Fisheries Management and Development was held in Rome, 27 June to 6 July 1984. It was attended by 62 ministers responsible for fisheries and delegates from 147 countries; by delegates from 3 liberation movements; by 14 representatives of the United Nations and its organs, organization, and bodies; and by observers from 24 intergovernmental organizations, and from 29 international nongovernmental organizations. The following is a summary of the strategy elaborated in the report of the conference published by FAO in 1984.

I. The contribution of fisheries to national economic, social and nutritional goals

(i) Countries have a sovereign right to determine their fishery policies. Fishery development plans should be an integral part of national economic development and food security plans and be in accord with social and nutritional goals.

(ii) Objectives should be based on assessments of the fishery resources available, existing technology, markets to be served, social and economic conditions, foreign operations, and other factors.

(iii) Multiple objectives are not necessarily compatible and priorities must be made clear.

(iv) Objectives must be reevaluated as the fisheries change.

(v) Governments must establish mechanisms and develop skills for planning.

(vi) Investment planning should follow scientifically based management measures, else be done judiciously.

(vii) National governments should collect data and statistics on all aspects of the fisheries and exchange them with neighboring countries.

(viii) Coastal countries must improve their abilities to assess stocks and determine allowable catches.

(ix) Development plans should include consideration of all technical, economic and social aspects.

(x) Administrators must work closely with scientists and businessmen.

(xi) Laws and government institutional structures are essential especially when competition exists for the fish and water resources.

(xii) Small-scale fisheries development needs close rapport between central and local governments and promotion of technologies adapted to local conditions.

(xiii) Governments should consider financial assistance to both small- and large-scale fleets.

(xiv) Development requires a balance between potentials and needs of both small- and large-scale fisheries.

(xv) The potential for freshwater fishery development should be considered.

(xvi) The expansion of fish farming should be encouraged in both marine and freshwaters.

(xvii) Plans must include protection of aquatic habitats threatened by degradation.

(xviii) Plans should include attention to development and marketing of low cost products for people.

(xix) Price policies must consider interests of both producers and consumers.

(xx) The potential for sport fishing should be considered.

II. Improved national self-reliance in fisheries management and development

(i) Training programs should be included in the national plans.

(ii) Training programs should be based on assessments of existing manpower and opportunities, including national, regional, or extraregional expertise.

(iii) External support and facilities should be sought particularly for training administrators and specialists.

(iv, v, vi) Arrangements should be provided for training high-level and mid-level administrators; technicians and scientists concerned with assessment, protection, exploitation and utilization of the resources; extension staff; and staff that can contribute to training.

(vii) Permanent regional training centers for stock assessment and evaluation should be established.

(viii) Special attention should be given to training staff.

(ix, x, xi) Efforts should be made to raise the educational level in fishing communities with attention to basic resource management, fishers organizations, and social development. Programs for women should be included.

(xii) Governments should try to ensure that trainees are effectively deployed and retained within the fisheries sector.

(xiii) Training programs should be regularly reviewed.

(xiv) Pilot projects should promote transfer of technology and use of infrastructure.

(xv) Consumers should be encouraged to eat fish products and processors should be trained to prepare them better.

III. Principles and practices for the rational management and optimum use of fish resources

(i) Governments should build up adequate research facilities, including research vessels, and formulate research programs.

(ii) The focus of management should be shifted toward entire ecosystems using the experience gained in the management of single stocks.

(iii) The government role should include collection and dissemination of information, formulation of objectives, adoption of policies, design and implementation of management measures, and continuous evaluation of results.

(iv) Effective fishery management mechanisms are necessary at all stages of development.

(v, vi, vii) Fishery management requires political decisions with which the fishers must be involved and which must be supported by legal and administrative frameworks.

(viii) Provisions for enforcement must be made.

(ix) Governments should ensure that fishers have clearly defined fishing rights but be restrained in order to avoid overcapitalization.

(x) Governments must control pollution and maintain critical coastal ecosystems such as mangroves.

(xi, xii) Neighboring countries should harmonize their management of shared stocks.

(xiii) International organizations should assist with surveillance of fishing in the EEZs.

(xiv to xix) Foreign fishing, allowed under permit, should be controlled to be compatible with local fishing. The foreign flag countries should cooperate by making available full information on their activities, by marking vessels appropriately, and by full compliance with coastal country laws and regulations. Neighboring coastal countries should cooperate on statistical, enforcement, and observer programs.

IV. The special role and needs of small-scale fisheries and rural fishing and fish farming communities

(i) The development of such fishing should be part of integrated rural development.

(ii) Optimum use of resources may require that economic and social development of small-scale fisheries be carefully coordinated with any development of large-scale scale fisheries.

(iii) Limited resources may prevent simultaneous improvement of both incomes and employment in small-scale fisheries.

(iv) Major constraints may be lack of infrastructure.

(v) Reduction of postharvest losses should be given high priority.

(vi) The link between extension services and practical research should be strengthened.

(vii) Active participation of the communities should be encouraged.

(viii) The role of women should be enhanced.

(ix) Fishers' organizations are important for consultation on decisions and as a channel for assistance.

(x) When necessary to reduce fishing effort, alternative employment should be found.

(xi) Local technology should be used if possible.

(xii) Special credit facilities may be important.

(xiii) Plans for water and land development should include consideration of the small-scale fisheries.

(xiv,xv) Better understanding of the social aspects of small-scale fishing is essential.

(xvi) Vessel safety is important.

(xvii) Training programs should consider best fishing techniques for the species available, and be adjusted to use of new equipment.

(xviii) Expansion of fish farming should be planned through regional cooperation.

V. International trade in fish and fishery products

(i) Improvement of market information is important.

(ii) Product quality must be raised to levels required by export markets.

(iii) Export operations should be directed toward final products rather than raw material, and toward diverse markets.

(iv) Trade barriers should be reviewed and modified.

(v, vi, vii, viii) Regional trade, especially among developing countries, should be promoted through trade preferences, technical advice, market information, harmonization of inspection procedures, and regular consultation.

(ix) All countries should help coordinate the activities related to trade promotion of the numerous international organizations.

VI. Investment in fisheries management and development

(i) Fishery management and development should be part of national priorities and development plans and should be better supported by international and regional funding agencies.

(ii, iii) Better information should be sought on the state of the resources, the technology under consideration, market demand and prices, training and extension needs, and socioeconomic matters, in order to improve the return on investment in fishery projects.

(iii) The time required for preparation of investment projects should be shortened.

(iv) Provisions should be made for continuing technical assistance which may be needed for many years.

(v) Loan programs should made more flexible through lines of credit for local relending institutions and arrangements for small loans to individuals.

(vi) The investment package should include provisions for institution building including the social investment needs.

(vii) National credit systems should be made more accessible to small-scale producers, processors, and distributors.

(viii) Governments should insure that investment in the fishery sector is comparable to other sectors of the economy and will attract private investment.

(ix, x) Technical assistance agencies and funding institutions need to cooperate more effectively.

(xi, xii) Continuing operation of investment projects should be enhanced through full attention to training and participation of personnel during project implementation.

VII. Economic and technical cooperation in fisheries management and development

(i) Cooperation among countries should be sought to enhance national programs.

(11) Joint ventures should include training of national personnel and transfer of technology.

(111) Special attention should be given to regional collection and exchange of data.

(iv) Regional fishery bodies provide a mechanism for sharing training and experiences among neighboring countries. Such collaboration should enhance national programs.

(v) Opportunities should be sought for further collaboration on joint fishing enterprises, new product technologies and development, and intraregional trade.

(vi) Business people and industry associations should be involved in program development.

(vii) The small island states need special efforts to improve their information base and establish common services.

(viii) The regional and subregional bodies need better financial support.

(ix) The negotiating capability of developing countries should be reinforced.

VIII. International cooperation in fisheries management and development

(i) Objectives of international cooperation should include contributions from all parties.

(ii) Cooperative research and cooperation through international bodies should be fully integrated with national plans and should make the fullest possible use of existing facilities.

(iii) Scientific and technical advice provided by international bodies should be made fully available to all concerned.

(iv) Better information on socioeconomic aspects of fisheries should be obtained.

(v) Central collection and analysis of data from all fleets fishing a common resource should be encouraged.

(vi) Additional support should be provided for regional bodies, and developing countries should increase their participation.

(vii, viii) Closer collaboration should be established among FAO regional bodies, other international organizations, and the several UN agencies.

(ix) More attention should be given to international cooperation with regard to the role of fish in nutrition.

(x) Cooperative use of research vessels should be sought.

Appendix B

Major References on Development

The preceeding discussion is an attempt to cover a very complex topic. The following are recommended to those who want to delve more deeply.

GENERAL

1. W. C. Baum and S. M. Tolbert, "Investing in Development: Lessons of World Bank Experience." Oxford Univ. Press, New York, 610 pp., 1985. This is the major condensation of development issues and concepts of the developing countries.
2. P. Dasgupta, A. Sen, and S. Marglin. "Guidelines for Project Evaluation." United Nations, New York, 383 pp., 1972. Includes methodology and applications.
3. E. Owens and R. Shaw, "Development Reconsidered: Bridging The Gap Between Government and People." Lexington Books, Heath and Co., Lexington, Massachusetts, 190 pp., 1974.
4. J. L. Simon and H. Kahn, "The Resourceful Earth." Blackwell, London, 585 pp., 1984. An analysis of trends and current condition of our resources.
5. S. Wortman and R. W. Cummings, Jr., "To Feed this World: The Challenge and the Strategy." Johns Hopkins Univ. Press., Baltimore, Maryland, 440 pp., 1978. Based on long experience with the International Agricultural Development Service (IADS).

6. UNDP, "Fisheries Development: Review of Support by the United Nations Development Programme." UNDP, New York, 48pp. + 9 tables and 18 appendices. Includes an evaluation of UNDP projects and future prospects.

FISHERY SCIENCE

7. E. E. Brown, "World Fish Farming: Cultivation and Economics. 2nd Ed. "Avi Publ. Co., Westport, Connecticut, 516 pp., 1983. A manual of methods.
8. J. A. Gulland, ed.,"Fish Population Dynamics." Wiley, London, 372 pp., 1977. A collection of fifteen articles on the market fisheries by the world's leading fishery scientists.
9. T. Panayotou, ed., "Small-Scale Fisheries in Asia: Socioeconomic Analysis and Policy." Internat. Dev. Res. Cen., Ottawa, Ontario, 283 pp., 1985. A collection of 28 papers comprising a major application of the fishery social sciences.
10. W. F. Royce, "Introduction to the Practice of Fishery Science." Academic Press, Orlando, 428 pp., 1984. A comprehensive introductory text.
11. R. R. Stickney, "Principles of Warm Water Aquaculture." Wiley, London, 375 pp., 1979. A major textbook and reference.

ECONOMICS

12. P. T. Bauer, "Equality, the Third World, and Economic Delusion." Harvard Univ. Press, Cambridge, 293 pp., 1981. A critical commentary on the development process.
13. J. P. Gittinger, "Economics of Agricultural Projects." Johns Hopkins Univ. Press, Baltimore, Maryland, 221 pp. 1972. Prepared for agriculturists--equally useful for aquaculturists.
14. R. M. Lawson, "Economics of Fishery Development." Praeger Publ., New York, 283 pp., 1984. A thorough coverage by an economist with extensive fishery experience in developing countries.

In addition to these major books, the long series of publications on fishery management and development by FAO and on general economic development by the World Bank should be consulted.

INDEX